W9-ALK-337

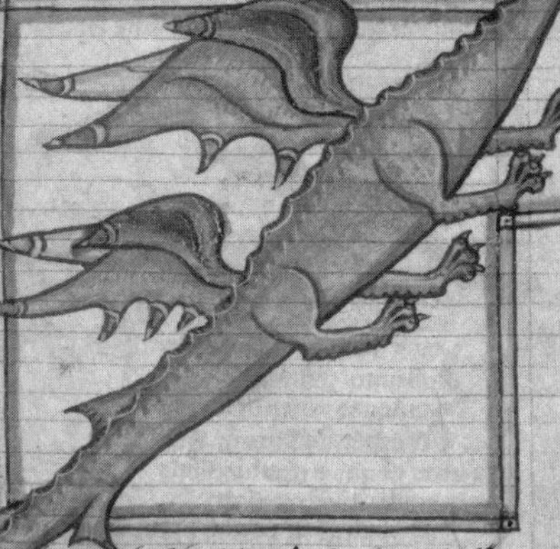

ALSO BY MICHAEL S. MALONE

The Future Arrived Yesterday: The Rise of the Protean Corporation and What It Means for You

Bill and Dave: How Hewlett and Packard Built the World's Greatest Company

The Valley of Heart's Delight: A Silicon Valley Notebook 1963–2001

Betting It All: The Entrepreneurs of Technology

The Big Score: The Billion-Dollar Story of Silicon Valley

THE GUARDIAN OF ALL THINGS

The Epic Story of Human Memory

Michael S. Malone

ST. MARTIN'S GRIFFIN

NEW YORK

www.stmartins.com

Frontispiece artwork courtesy of the British Library

The Library of Congress has cataloged the hardcover edition as follows:

Malone, Michael S. (Michael Shawn), 1954–

The guardian of all things : the epic story of human memory / Michael S. Malone. — 1st ed.

p. cm.

Includes bibliographical references and index.

ISBN 978-0-312-62031-8 (hardcover)

ISBN 978-1-250-01492-4 (e-book)

1. Memory. 2. Technology and civilization, 3. Civilization—History. I. Title.

BF371.M3395 2012

153.1'2—dc23

2012010246

ISBN 978-1-250-02323-0 (trade paperback)

St. Martin's Griffin books may be purchased for educational, business, or promotional use. For information on bulk purchases, please contact Macmillan Corporate and Premium Sales Department at 1-800-221-7945, extension 5442, or write specialmarkets@macmillan.com.

First St. Martin's Griffin Edition: December 2013

10 9 8 7 6 5 4 3 2 1

Contents

Introduction
The Guardian of All Things
Memory as Biography

"Memoria est thesaurus omnium rerum e custos."
Memory is the treasury and guardian of all things.
—MARCUS TULLIUS CICERO

I first conceived of writing about the story of memory more than thirty years ago, as the result of two experiences in my early adulthood.

I joined the *San Jose Mercury News* as a daily technology reporter—probably the first in the world—in late 1979. It was a thrilling beat: Silicon Valley was in its first great flowering, and other than a few trade reporters, I had it all to myself. I covered the last great days of Hewlett-Packard and the first few years of Apple. I interviewed all of the soon-legendary first generation of Valley entrepreneurs, from the men who founded the semiconductor industry to the builders of the first personal computers, disk drives, and video games.

I felt myself (accurately, as it turned out) to be uniquely positioned to watch up close the most important story of my time as it unfolded. And as I raced around the Valley and did my interviews and wrote my stories I learned two things. One was that even in as scientific and empirical an industry as high tech, the real story was always elusive and the official histories were rarely correct.

But it was the other thing I learned that eventually led me to write this book. Wherever I went in those days, I was told how lucky I was to be covering Silicon Valley at that moment. That the current efflorescence of new companies and new products would be a brief one. And that the glory days were already coming to an end.

Why? *Memory*, they told me. Chips and processors are getting faster and more powerful by the year. But they are solid-state; they need only be miniaturized; their path is clear. But memory, I was told, is a different story: Chip memory can't keep up the pace, and magnetic memory is all about spinning disks and moving heads and motors. Disk memory is a machine with moving parts; it'll never keep pace with logic chips and microprocessors.

And yet as the years passed, almost magically, memory kept up. Somehow, the engineers and scientists who worked in memory always found a way to progress each of the technologies—semiconductor, magnetic, optical, and odd little experimental designs like magnetic "bubbles"—and always managed to keep up. Sometimes these companies, despite often being in competition with one another, seem to work together—like rugby players making their way downfield, flinging lateral passes from one company or industry to the next . . . always making forward progress, and always staying just ahead of the other guys.

It struck me then, and I still believe, that these memory folks were the great unsung heroes of the electronics age. While others enjoyed the limelight and the glory for their achievements, those who toiled in the memory business—on whom all of the others depended for survival—did so (with the rare exception of an Al Shugart) in the shadows of near-anonymity. I resolved then to tell their story.

The second experience was the death of my father.

My dad was a remarkable character: tough, brilliant, and a hero. When he died at sixty-seven, he had lived what for most men would have been a couple lives. He had survived swallowing razors in a carnival sideshow, murderous hobos riding the rails, thirty missions as a bombardier in a B-17 over Europe, gun battles as a spy in Germany and North Africa, encephalitis, car crashes, a Tibetan earthquake, and three heart attacks. And in the end, what killed him was that most common domestic injury of middle-aged men: He fell off a ladder and fractured his skull.

He didn't die quickly; he was too tough for that. Instead he lingered, in a semicoma, as his wife and adult children visited him each day in the intensive-care ward of the county hospital.

It was on one of those days when my wife and I visited him that we were met in the entryway to the ward by one of the doctors. "Your father," he told me, "is doing the most unusual thing. We can't figure it out. Would you watch him for a while and tell us what you think it is?"

I pulled a chair up next to my father's bed and leaned close. At first, he looked as he did on all of the other days: stretched out in his slightly tilted hospital bed, a bank of monitors beside him, and an oxygen mask on his face. But something had changed: Instead of lying quiet and still, my father's eyes were open and unfocused. His left hand, all but immobile until now, had crept up above the top of the blanket, and his fingers were tapping the air. And beneath his oxygen mask, I could see my dad's lips moving, as if he were speaking in a crisp and systematic way.

I leaned over and put my ear to his cheek. I could see his fingers still tapping the air just in front of me.

I heard the doctor's voice behind me. "Do you know what he's doing?"

I stood up and turned around. "Yes. I believe he's going through the preflight checklist for a B-17 bomber. The oxygen mask must have set off a memory. See his fingers? I think he's flipping toggles and switches and tapping gauges."

"Really," said the doctor. "Amazing." And he walked away.

I watched my father. Somehow, a half-century later, it was all still there—every step, every reading, and every switch to flip. I'm certain that even he didn't know he remembered. After all, he had washed out of pilot school—that's why he became a bombardier and navigator. So his time in the pilot's seat must have been a brief one. And yet, that had been enough to imprint the memory of an incredibly complex experience, at age twenty-two, into my father's brain until the day he died. What else was in there? Was *everything* in there—all of the memories of a lifetime? And were they in all of us?

I have always been haunted by what I saw in my father that day. Losing him was painful enough—but the realization that his whole lifetime of memories, still present, might have been lost as well added to the heartbreak. And in the years since, I've followed the latest discoveries in the cognitive sciences and neuroscience, stories of prodigious feats of human memory, and the growing body of work in brain implants and brain-machine interconnections. All of them hinted, as did my father that day, of something vast and awesome hidden just beyond our reach. So did watching my two boys, as babies, as they seemed to grab the entire world with their minds.

My own career has been a busy one, not least the writing of a small shelf full of books bearing my name as author. Through it all, I never gave up on my plan to write the story of memory, both human and artificial, and how they interact to make us who we are. Several times over the years I proposed the book to publishers . . . only to have them (figuratively) yawn at the prospect of a dreary tome about disk drives, and counterpropose some other book idea ("How about a book about Apple?"). About a decade ago, I bounced the idea back and forth with the BBC's former NASA reporter and legendary science historian James Burke (author of *Connections*), an old friend. But our plans went nowhere—though his readable style and sense of fun is something I've aspired to in this book.

One thing I've learned in a long career as a journalist and author is that if you wait long enough, people will come around to even the craziest idea. In the end, it was an ABCNews.com column of mine, no doubt combined with the popularity of books like Ray Kurzweil's *The Singularity Is Near*, and the ubiquitous little miracle of the semiconductor memory "thumb drive," that at last made memory memorable . . . and after twenty years of trying, I was finally given the opportunity by the kind people at St. Martin's Press to write this book.

Strangely, after decades of patiently waiting for the chance, when I finally sat down to write it, I felt under the gun—as if I were in my own race against the future. Because of the technological innovations of the intervening years, what I had envisioned as a celebration of the magnificence of the human brain and the superhuman achievements of the builders of artificial memory had now become something more ominous and compelling. There was a growing sense in high-tech, scientific research, computer science, and even philosophy and religion that the two great paths of memory, the one internal and organic and the other external and mechanical, having run in parallel now for millennia, were suddenly about to converge—perhaps even in our lifetime. And when that occurred, the collision might very well change the very meaning of being human. Suddenly I was not just an entertainer but a messenger; not just a historian but a futurist.

That's not to say I didn't enjoy myself writing this book. There is nothing like sweeping across the entire span of human existence, drawing connections between events thousands of years apart, bringing to their proper prominence events that are normally ignored by history texts, and showing how a decision made *then* has, through a long causal chain of subsequent decisions made by the most diverse imaginable collection of men and women, become an imperative in our lives *now*.

But most of all, in the final chapters, I got to visit with a lot of old friends and acquaintances, many of them now long gone, who created our fast-moving, scary, and often exhilarating high-tech modern world. To see them now as honored historical eminences—and to remember them as just hardworking, everyday men and women—was to be reminded that both genius and glory are within the reach of everyone if we're willing to risk everything in pursuit of our vision. I'm thrilled that many of them are still around to see their story told at last.

As you will see in the pages ahead, the story of memory is, in the end, the story of freedom. From Paleolithic tribes to modern nations, whoever controls memory holds power. So the history of memory—of the evolution of the human brain, of the invention of speaking and writing, of the invention of ever-new ways to record experience, of the mastery of machines and the exploration of mind, of the slow extension of ownership of memory from shamans to rulers to scribes to bureaucrats to everyone—is also the history of the liberation of the human spirit. It is a memoir of memory, and of the democratization of remembrance—and its control.

Each chapter of this book is the story of one major step in the history of memory. Each has its protagonists, its paradigm-shifting new invention, its cultural transformation . . . and, in the end, its new hunger created by the fulfillment of the last desire. That hunger will never disappear; it was with us when Heidelberg Man struggled to make himself understood 500,000 years ago . . . and it will be with us on the day, if it comes, when we decide to shed our skins and, in search of immortality, climb into our machines.

I began this book thinking of memory as a discrete phenomenon: a region in our brains and a feature in our computers. I finished this book understanding that both are inextricable from what they serve. To lose our memories, as human beings, is to lose our identity, just as the loss of memory to a computer reduces it to little more than an adding machine. Good or bad, memories provide us not only with identity, but existence itself—and to forget, or to be forgotten, is its own kind of death.

As I said, when I began to write this book I thought, after twenty years of waiting, I could now take my time. And certainly there was no need to hurry when writing about Gilgamesh, or Saint Isidore, or even Giordano Bruno. But as I delved deeper into the recent works of scientists I knew, such as Ray Kurzweil and Gordon Bell, it struck me that the next chapter in the story of memory that they were predicting—if accurate—was so shocking and so imminent that it might crush many of us in its path. What had become a lark to tell became a duty to warn . . . even if I personally doubted the likelihood of those events occurring.

Finally, if you haven't concluded this already, I believe that the story of memory is as epic, heroic, and thrilling as any comparable story of battles, empires, artists, or monarchies across the span of human history. And I hope that I have captured at least a little of that in the pages that follow. And I hope I've honored the men and women of memory whose achievements I promised to celebrate all those years ago.

Comedy or tragedy, in the theater of memory we are all players. So, raise the curtains and we'll begin with the first act.

THE GUARDIAN OF ALL THINGS

I
Finding a Voice
Memory as Word

When did hominids become human?

When we used our memories to do more than remember.

When did that occur? That's not any easy question to answer—certainly not as easy as it might seem from those charts of the "Ascent of Man" we first encountered as children and have seen parodied ever since.

There, the answer was obvious: in the long parade from the knuckle-dragging *Ardipithecus ramidus* to the upright *Australopithecus* to *Homo habilis* carrying his stone scraper, to *Homo erectus* with his flint knife. Next, depending upon the complexity of the chart came Neanderthal Man, carrying a spear on his sturdy shoulders.

Finally, and you can tell this was the culmination of human evolution because the figure was standing upright, his hair combed, looking European, and in some posters even clothed to protect our same-species modesty, stood *Homo sapiens* in all of his mid-twentieth-century glory. He always looked about to shake your hand and introduce himself as the Southeastern regional sales manager.

Implicit in the chart was the notion that man became Man somewhere in midstride between heavy-browed Neanderthal and shiny new Brussels bureaucrat. Older charts made things a little more complicated, because, unexpectedly, there was a new guy (also covered in skins because he looked a little too much like one of your neighbors) slipped in between Mr. Forehead and ourselves. This was Cro-Magnon Man, who—until he was judged to be just modern man in mufti—presented the depressing paradox of apparently being a more impressive specimen than modern man.

Still, missing links (real or imagined) or not, the Ascent of Man seemed a neat and straight parade across the course of a couple million years. And the threshold between early man (simple tools, hunting parties, small family groups) and modern man (computers, cities, nation-states) lay somewhere in that blank space just in front of Mr. Neanderthal's mighty forehead.

Part of the appeal of limited knowledge is that it often makes organizational schemes and taxonomies really easy. But eventually you dig up enough bones and fossils that you have to put a new head—and a new name (Apatosaurus)—on everyone's beloved *Brontosaurus*. And so, in the four decades since the discovery of "Lucy," the little female *Australopithecus afarensis* who lived more than three million years ago, the Ascent of Man has changed from a single-file march through history into something closer to the crowd at the end of a hockey game milling around in the plaza and slowly making its way to a single exit turnstile.

Every year, archaeologists, armed with ever more powerful investigative tools, find new bits of bone and other artifacts that alter—sometimes radically—our understanding of hominid history. And it is only going to get worse: Thanks to phenomena such as "genetic drift" (in which rare genes can come to dominate isolated populations) the closer we look, the more alternative evolutionary pathways, breakthroughs, and dead ends we are likely to find.

A case in point was the controversial discovery in 2004 on the Indonesian island of Flores of the tiny bones of "Hobbit people" (*Homo floresiensis*), most of them less than three feet tall.[1] Remarkably, some of these skeletons are only about 13,000 years old, making them contemporaneous with modern man. Whether the Hobbit people were a distinct species or merely the bones of *Homo sapiens* with genetic diseases (such as lack of a thyroid gland) is still being debated.

But perhaps the most interesting recent discovery comes from the spot just *behind* Neanderthal Man's huge head on our old chart. His name is *Homo heidelbergensis* (named after the university) and though first identified at the beginning of the twentieth century, his importance wasn't really understood until the 1990s—which is why he is all but unknown to the general public. *H. heidelbergensis* both answers and complicates an important question in human evolution: What happened during that apparent transformation from Neanderthal to modern man 50,000 years ago in Asia (30,000 years ago in Europe)?

The answer, scientists now believe, begins with *H. heidelbergensis,* who appeared about 600,000 years ago. *H. heidelbergensis* was an impressive figure: heavily muscled, six feet tall (there may even have been some seven-footers), and with a brain about the size of modern man's. He knew how to use simple tools. And, most remarkably, he also appears to be the ancestor of *both* Neanderthal and modern man.

That helps to explain why, at least toward the end of the former's existence, Neanderthal and modern man appear to have existed side by side. Neanderthal Man got most of Heidelberg Man's looks—the heavy bones, beadle brow, and the comparatively large brain. But modern man got the height and added a uniquely flat face and a lanky frame. And if modern man didn't inherit quite the cranial size of his Neanderthal cousin, he instead got something even more important: language, and a brain to process and store it.

As anyone who watches science documentaries knows, chimpanzees and apes, for all of their brainpower (our appreciation of which also seems to grow by the day), have limited ability to speak because of a weak larynx due to a narrow cervical vertebra. What most people don't know is that, as with deaf-mutes, this inability to verbalize is further constrained by an inability to *hear*—in the ape's case, a lack of capacity to differentiate between certain vocal sounds.

Of all of the points of divergence between man and apes, this is a big one—a fact made clear when researchers first taught apes to use sign language and were stunned by their facility. Historically, this divergence appears to have taken place with Heidelberg Man's immediate predecessor, *Homo ergaster,* a southern African hominid who should also stand on the chart between *H. habilis* and Neanderthal Man.

Like Heidelberg Man, *Homo ergaster* is a very interesting character. Huge—he may have averaged a couple inches taller than six feet, and the females were nearly as tall—*Homo ergaster* appears to have used his comparatively larger brain not only to gain mastery over fire but to take hominids out of Africa for the first time. As earthshaking as those achievements were, *H. ergaster*'s greatest achievement was to evolve both a wider cervical vertebra—which gave him the first "human voice"—but also a new middle- and outer-ear configuration that enabled him to hear the voices of others.

This was hardly a coincidence. The competitive advantage of a more facile voice was amplified by the improved hearing of that voice—and vice versa. *H. ergaster* probably never had a true speaking voice, much less developed a spoken language; his brain was still too small for that. But as with fire and tools, *H. ergaster* was, if not the ultimate owner of spoken language, certainly its pioneer.

The task of turning this capability into a defining human characteristic fell to *H. heidelbergensis*, with his larger and nimbler brain. Certainly he had the physical tools to do so. And there is a wealth of circumstantial evidence—for example, he was the first to honor and bury his dead, he developed relatively precise tools, and may have collected red ochre for painting and body adornment—that suggests a level of cultural sophistication that seems impossible without some kind of complex form of communication.

But was it a spoken language? We may never know the answer to that. There are no known Heidelberg Man drawings or carvings to suggest pictogram-based communication. And as anyone who has ever been in a hunting party or a reconnaissance team knows, it is possible to convey a considerable amount of information—even without a formal grammar—via a very small repertoire of hand signals. Indeed, being prodigious hunters operating in small family units, Heidelberg Man may well have created a kind of sign language like that found in later hunter-gatherer societies (such as American Indians) to create a kind of lingua franca for those rare intertribal encounters.

But that larynx and inner ear weren't evolving without a competitive advantage. So, we can assume that Heidelberg men and women were communicating with one another with an increasingly sophisticated vocabulary of sounds, if not yet a true language. Further, we can also assume that these sounds were taught to thousands of generations of progeny, who slowly but surely added to the common repertoire. And, given the flexibility with which these sounds could be made by the evolving voice box—and the fact that verbal communication didn't have to be line-of-sight—it seems pretty likely that, had you been walking in the Bavarian woods a half-million years ago, you would have heard proto-humans calling out to one another across the valleys and through the forests.

Spoken language might not have conferred much of an additional advantage during the hunt, but it certainly did before with strategy and after with the distribution of the spoils. And in a world of terrors and dangers, shouted warnings would be especially useful—especially to warn people looking the wrong way, or to assemble the tribe quickly in an emergency.

But this is as far as Heidelberg Man got, even in the most optimistic analysis. He could convey information (*look out!*) in the present—and perhaps reference physical objects in the past (*the wooly mammoth with the crooked tusk we killed*)—but not much more; and even the latter was probably better expressed with sign language.

More sophisticated spoken language would have to wait for Heidelberg's descendant, Neanderthal Man. This may come as a surprise to many readers, who were taught in school, movies, and television that Mr. Neanderthal, the stereotypical "caveman" only managed to communicate with an array of grunts and gutturals. But anthropologists have challenged that notion ever since 1983, when an Israeli dig of Neanderthal skeletons uncovered a hyoid bone.[2] The hyoid bone is a *c*-shaped structure that acts like a roof truss, tying together the tongue and the larynx and enabling them to brace off each other to produce a wider spectrum of sounds.

Hyoid bones have been around a very long time—they evolved from the second gill of early fishes—but the particular shape of these Neanderthal hyoids was thought to be unique to humans, creating a "descended" larynx that enables *Homo sapiens* to not only wrap their voices around an endless array of sounds but also to sing notes across multiple octaves. Now, it was found, Neanderthal had the same hyoid bone. Genetic research also found that Neanderthal shared with modern man the FOXP2 (Forkhead Box Protein P2) gene form equated with language capability.[3] These discoveries have led some researchers to speculate that Neanderthal men and women, like their modern counterparts, may have verbally communicated in two forms: speech and music.

A SIMIAN SONG

A singing Neanderthal is a long ways from the classic image of the slope-headed caveman in a bearskin dragging a club. But for all of his heavy features and brute strength, Neanderthal man was a very sophisticated, even artistic, creature, with a brain larger than (if not as advanced as) modern man's. He built shelters and fabricated fine bone and antler tools, buried his dead, lived in extended communities, and organized some of the bravest and most sophisticated hunts in hominid history. Given all of that, it would not be surprising that Neanderthal Man, in some limited way, actually talked.

One archaeologist, Steven Mithen of the University of Reading, has argued that Neanderthal Man used a "protolinguistic" mode that reflected the fact that spoken language and music had not yet split and taken their separate paths.[4] This simple, singing language he calls *hmmmm*—meaning that it was holistic, manipulative, multimodal, musical, and mimetic. But at the moment, Mithen's a lonely voice: The current scientific consensus is that a language of even this little sophistication was still probably too much for Neanderthal Man.

Temporarily leaving the disturbing image of heavy-browed Neanderthal hunters serenading one another with song as they maneuver through the snow around a trapped wooly mammoth, let's pause to consider what was going on inside the brain of one of those hunters.

First of all, put aside any old stereotype about our Neanderthal Man being stupid. He had a very powerful brain; in some ways—pattern recognition, multisensory processing, and comprehensive visual field analysis—it was likely more powerful than our own. But a modern human being would find inhabiting such a mind an enthralling—and terrifying—experience indeed.

What would be exciting about sharing Neanderthal Man's mind is that it would exist in the present with an intensity few of us have ever known. The world around us would explode with so much information—sounds, colors, movements, shapes, textures, and smells—that it would almost be overwhelming. We modern humans consume drugs, watch movies, seek out adventures, and take dangerous risks just to feel for a few moments an eternal here-and-now that Neanderthal Man likely felt almost every second of the day.

That's the good news. The bad news is that the cost of so completely owning the present is to lose all of the future and much of the past. Finding yourself inside a Neanderthal brain would be, in those intervals when you weren't absorbed in the moment, a desperately lonely place. For one thing, that voice—of consciousness and conscience—that we hear in our heads would be gone. So would all of the stories and anecdotes that we've ever heard, and every memorable conversation we've ever had. What things we did learn—almost always by observing someone else—would be learned by rote and, because we lacked any ability to analogize, would be difficult to adapt to changed circumstances. Instead we would just ritually do things over and over, not understanding why it no longer worked.

Without a real language, our ability to interact with others would be severely limited. Certainly a lot can be accomplished nonverbally: hunting, child-rearing, sex, food preparation, and so on. But the inability to share fading memories of the past, or dreams of the future, of one's hopes and fears, of new ideas and useful experiences . . . that would be devastating to anyone who had ever known such things.

But ultimately, what might be the most frustrating thing about finding ourselves in a Neanderthal brain would be that despite our new hyperacute sensory experience of the world around us, something profound and vital would be missing from our ability to *enjoy* those experiences. Not only would it be difficult to share those experiences with others but, just as important, we would have almost no capacity to contextualize them to ourselves. Without language, we lack a capacity for analogy and metaphor—perhaps the most important traits distinguishing human beings from all other living things on Earth. Without those two, our capacity to learn and grow intellectually would be profoundly impeded.

Neanderthal memory was likely a powerful engine, capable of remembering an almost infinite number of warnings of impending weather change, animal patterns and behaviors, precise geological and biological features along long migratory paths, and the tracks, spoor, and calls of hundreds of species and varieties. But without the ability to extrapolate from limited data to larger explanations, or to tie together disparate subjects in order to increase understanding—and, most of all, to tie the trends of the present into scenarios for the future—Neanderthal Man was doomed to be forever trapped in the "now," in his short and brutal lifetime having to learn almost everything from experience.

Neanderthal Man had a human brain, but without language—or even with a protolinguistic like hmmmm—he couldn't operate it efficiently. Most of all, without language, he couldn't properly fill, organize, expand, or access his memory beyond a direct correlation between past experience and current stimulation. And without such a memory, Neanderthal Man couldn't be fully human.

But 200,000 years ago, a second descendant of *H. ergaster* emerged in Africa. This hominid would take another 150,000 years to reach its ultimate form and on at least one occasion (the Toba Catastrophe 70,000 years ago) would come so close to extinction that the entire population might have fit in a couple Boeing 747s.[5] But this creature, *Homo sapiens*—modern man—survived. We typically, and admiringly, credit this to mankind's vaunted tenacity and will to live. True enough, but in the end what may have saved us (and enabled us to quickly recover) was our ability to work together in common purpose, to preserve our acquired knowledge and to build one innovation on another to create a kind of cultural momentum.

Archaic Homo sapiens, as this earliest form of modern man is called (the Cro-Magnon Man of the old charts) seems, with his skinny frame, light muscles, and delicate skull, awfully frail compared to his more robust predecessors. Indeed, it's hard to see how he could have survived in the late Ice Age world.

But against all odds, he did. And the reason goes beyond anatomy. In fact, maybe for the first time in the 2 billion years of life on Earth, *Homo sapiens* men and women succeeded because they had found a way to make their physical attributes only secondary in importance. And they did so precisely by evolving those specific features that had appeared in their ancestors—a wide cervical vertebra, an advanced hyoid bone, an improved inner ear, and the FOXP2 gene—and then added one more new physical feature: a prominent and pointed chin that made possible a number of new verbal sounds, including clicking.

But the real advances were made inside the skull of these first modern men. All of this improved communications equipment meant that early modern man not only could both speak and hear increasingly complex utterances but could attach a vast array of different sounds—first phonemes, then syllables, then words, then sentences—to things and events in the natural world.

In other words, early modern man could *talk*. And in the construction of those words—that is, in the attachment of multiple sounds to increasingly complex phenomena—human language slowly evolved from the concrete to the abstract, from direct correspondence between object and name into the messy, inexact, and infinitely powerful world of analogy and metaphor.

Needless to say, accomplishing this required a whole different kind of memory—one dedicated as much to organization and filing as mere storage—that was as much language-based as stimulus-based, and was as much optimized for synthesizing new concepts from disparate pieces as it was for speed of access.

By turning memory from mere storage into a language-based scheme of organization, early modern man solved a puzzle that had challenged its ancestors for more than a million years: *How do you use that powerful brain to transfer complex information to others?* And in finding that solution, early modern man had discovered a remarkable side effect: That same facility with language also created a much richer internal life of the mind, filling it with stunning relationships, stories, and astonishing thoughts about life and death; about the Earth and the cosmos; and even about God and man's relationship to the Divine—thoughts that no living thing had ever thought before.

And with such thoughts, these early *Homo sapiens* (which translates as "knowing men") became human.

TOTAL RECALL

Memory is almost as old as life itself.

All animals have some form of memory, even if it is only a simple biochemical encoding of aversion to negative stimuli. Even schoolchildren know that a paramecium, once it has been adversely stimulated as a result of a certain type of behavior—for example, being electrically shocked for moving in a certain direction—will quickly "learn" not to act that way again.

What this means is that while you may need some kind of brain to think, you don't necessarily need one to "remember." All that's really needed are some organic chemicals, which work similarly to an electronic resistor—linked to some physical action or behavior that can register environmental change. Thus, the "memory" of the unpleasant event is encoded in the chemical and results in a change in future response.[6]

Of course, it becomes a whole lot easier when you have many cells, and some of them—neurons—are dedicated to the job of sensing these changes, converting them into fast-reacting chemicals or electric charges, and transferring that information to other nerve cells in the body to create both a rapid response and an enduring record of what just happened. That's essentially how the first animals functioned a billion years ago—and how the simplest ones do today.

The next step is to bundle those neurons into cords—nerves—that are organized like the roads and highways of a modern state. That is, you use small roads, streets, and byways to reach almost every point on the landscape. Then you merge them into higher-traffic arterials that eventually become on-ramps of highways and then a great interstate superhighway. This process took another half-billion years and resulted in increasingly sophisticated animals called chordates, a group that includes both human beings and tapeworms, and almost every crawling, swimming, walking, or flying creature in between. Chordates share a "peripheral" nervous system with their more primitive cousins, but also feature a distinguishing "central" nervous system that, in the simplest chordates such as flatworms and flukes, is little more than a proto–spinal cord.

As animals continued to evolve, the more advanced chordates not only began to protect their central nervous systems with bone and cartilage—spinal columns or backbones—but also to feature two distinct types of neurons: *sensory*, or those that send signals to the spinal cord; and *motor*, which send the response commands back to muscles and organs. This is basically the nervous-system setup you see in insects, lobsters, and worms—proving that you don't need a particularly advanced nervous system to be fast, efficient, and in the case of a species like the cockroach, so perfectly evolved as to be almost immortal.

Once again, bugs and other creepy-crawlies have rudimentary memories. It's enough to bring them into the world with a complement of instinctual behavior that makes them formidable hunters, consumers, and adversaries . . . but not enough to keep them from returning to that same hot lightbulb until they cook themselves. But the components are now there; all that's needed is to perfect their configuration. One of the best solutions turned out to be putting a few knots in that central nervous system. With these bundles—"brains"—animals could begin to divide up tasks and store ever greater amounts of information. Tie a few of these knots along the course of a spinal cord—as in, say, a spider—and suddenly you've conferred some real competitive advantages in terms of the stalking skills and lightning response needed to be a successful predator.

It turns out, however, that evolutionarily speaking, the most efficient layout of the animal nervous system—especially if you want to tackle more complex challenges like migration, nest building, and baby rearing—is to combine all of those small bundles into one big brain. The milliseconds that you may lose in response time as the messages now have to travel the length of the spinal cord are more than made up for by the network effects of having all of those neurons in close proximity.

This need becomes especially acute when you start developing really efficient vision and a sense of smell, both of which consume a whole lot of brain processing power. In amphibians, and especially reptiles, these senses take up most of the brain. And indeed, they are so important that these animals began the process of producing a new kind of brain structure, the *cerebrum*—what we colloquially call "gray matter"—that contains neurons specifically organized to manage voluntary movements of the body, the senses, *language*, *learning*, and *memory*.

You can see where this is going. As animal life evolved—adding greater intelligence to existing orders and phyla, as well as producing whole new kinds of creatures such as birds and mammals with even greater intelligence—the new brains featured ever more gray matter. These growing cerebra likely were the result of the competitive advantage of having better visual perception and an improved sense of smell. But they carried in train an equally greater aptitude for facility with language, an ability to use that language to gain knowledge, and, not least, to store that knowledge using a language- and image-based filing system to improve access and recovery.

Exactly how all this worked has been a puzzle. Scientists long ago recognized that there are two distinct memory operations: *short-term*, which (as every cramming college student knows) seems to hold memories for about thirty seconds before they fade; and *long-term*, which seems to be a function of either a strong experience or repetition, and can last an entire lifetime. But how they work and what makes them different has long been a mystery. Only in the twenty-first century have researchers begun, using genetic manipulation, to get a glimmering of how memories are formed, erased, or stored.

As long suspected, memory creation is the result of a biochemical reaction that takes place in nerve cells, especially those related to the senses. Recent research suggests that short-term, or "working," memory operates at a number of different locations around the brain, with special tasks tending to be handled in the right hemisphere of the brain and verbal and object-oriented tasks in the left. Beyond that, the nature of this distribution, retrieval, and management is the subject of considerable speculation.

One popular theory holds that short-term memory consists of four "slave" systems. The first is *phonological*, for sound and language that (when its contents begin to fade) buys extra time through a second slave system. This second operation is a *continuous rehearsal system*—like when you repeat a phone number you've just heard as you run to the other room for your phone. The third system is a *visuo-spatial sketch pad* that, as the name suggests, stores visual information and mental maps. Finally, the fourth (and most recently discovered) slave is an *episodic buffer* that gathers all of the diverse information in from the other slaves, and maybe other information from elsewhere, and integrates them together into what might be described as a multimedia memory.[7]

Other theories hold that short-term memory is, in fundamental ways, just a variant of long-term memory. But almost all brain scientists agree that the defining characteristic of short-term memory is its limited functionality—both in duration and capacity. Simply put, short-term memory fills up fast—scientists speak of four to seven "chunks" of information such as words or numbers, which short-term memory can hold at any one time—after which its contents either fade or they are purged.[8] Rehearsal can temporarily keep important short-term memories alive, but ultimately the information must either be transferred to long-term memory or lost.

Long-term memory, though it uses the same neurons as short-term memory, is, as one might imagine, quite different in the use of those neurons. Whereas the short-term memory is limited in scope and capacity, long-term memory takes up much of the landscape of the upper brain and is designed to maintain a permanent record. Only in the last few years have researchers determined that memories are often stored in the same neurons that first received the stimulus. That they discovered this by tracking storage of memories in mice created by *fear* suggests that evolution found this emotion to be a very valuable attribute in a scary world.

Chemically, we have a pretty good idea how memories are encoded and retained in brain neurons. As with short-term memory, the storage of information is made possible by the synthesis of certain proteins in the cell. What differentiates long-term memory in neurons is that frequent repetition of signals causes magnesium to be released—which opens the door for the attachment of calcium, which in turn makes the record stable and permanent. But as we all know from experience, memory can still fade over time. For that, the brain has a chemical process called *long-term potentiation* that regularly enhances the strength of the connections (synapses) between the neurons and creates an enzyme protein that also strengthens the signal—in other words, the memory—inside the neuron.[9]

Architecturally, the organization of memory in the brain is a lot more slippery to get one's hands around (so to speak); different perspectives all seem to deliver useful insights. For example, one popular way to look at brain memory is to see it as taking two forms: *explicit* and *implicit*. Explicit, or "declarative," memory is all the information in our brains that we can consciously bring to the surface. Curiously, despite its huge importance in making us human, we don't really know where this memory is located. Scientists have, however, divided explicit memory into two forms: *episodic*, or memories that occurred at a specific point in time; and *semantic*, or understandings (via science, technology, experience, and so on) of how the world works.[10]

Implicit, or "procedural" memory, on the other hand, stores skills and memories of how to physically function in the natural world. Holding a fork, driving a car, getting dressed—and, most famously, riding a bicycle—are all nuanced activities that modern humans do without really giving them much thought; and they are skills, in all their complexity, that we can call up and perform decades after last using them.

But that's only one way of looking at long-term memory. There is also *emotional* memory, which seems to catalog memories based upon the power of the emotions they evoke. Is this a special memory search function of the brain? Is it a characteristic of both explicit and implicit memory? Or, rather, does it encompass both? And what of *prospective* memory—that ability human beings have to "remember to remember" some future act? Just a few years ago, researchers further discovered that some brain neurons can act like a clock in the brain, serving as a metronome that orchestrates the pace of operations for the billions of nerve cells there.

Why? These and other features are but a few of the conundrums in the long list of questions about the human brain and memory. What we do know is that—a quarter-million years after mankind inherited this remarkable organ called the brain—even with all of the tools available to modern science, human memory remains a stunning enigma.

A THIN LAYER OF THOUGHT

With the rise of the hominids three million years ago nature found two valuable tricks to aid brain development. The first was to divide the brain's activities into different regions, with the autonomic nervous system (that is, the parts of the body that run on their own, such as the heart, lungs and the many glands) managed by the lower, older parts of the brain, and the somatic (or voluntary) nervous system handled by the gray matter of the cerebrum. The higher functions of the latter, such as speaking and fine-detail motor skills, found their home in the ever-growing frontal lobes of the hominid brain.

But Nature saved her biggest trick for last: the *cerebral cortex*. This was a final layer of neurons—what might be called "very gray" matter because when it is preserved it is darker than the whiter neurons beneath—covering the entire surface of the upper brain. Untucked from the many brain folds that are designed to maximize its size in the confined space of the skull, the cerebral cortex is basically a four-millimeter-thick sheet of 10 billion neurons covering about 1.3 square feet.

Wadding a sheet of neurons that size into the comparatively small space of the *Homo sapiens* skull proved to be a huge challenge to evolution—one fraught with numerous vulnerabilities, but one with enough adaptive advantages to make those risks worthwhile.

And those risks were considerable. To carry that giant-sized brain, modern man lived with a thinner skull and less cranial padding than his Neanderthal counterpart, increasing the likelihood of concussion and fatal head trauma. It also led to newborn human babies with skulls sized at the very limit of the adult human female pelvis . . . making childbirth, until only the last hundred years in the developed world, the leading cause of death for young women.

But in terms of adaptation and the survival of the species, the cerebral cortex was worth the cost. That's because within those billions of neurons was a capacity for integrating sensory input, abstract thought, language, and a prodigious memory that had no equal—indeed, almost no precedent—in the history of life on Earth. And incredibly, that was the least of it, because somehow, in a process that remains inexplicable, from this sheet of nerve cells emerged that most extraordinary and singular of traits: *consciousness.*

Someday, the odds (and the Drake equation) suggest, we will find life elsewhere in the universe.[11] And if that happens, almost infinitesimally smaller odds predict that we will also eventually meet another conscious life form. But for now, and maybe for thousands of years to come, we *Homo sapiens* alone are both conscious and self-reflexive. For now, it seems that only we can appreciate that loneliness, and with our imaginations cast our minds to the edge of the cosmos in search of answers. Only we know that we are *we.*

Whether you believe in a divine spark, a network effect emerging from those billions of neurons, or some kind of quantum phenomenon taking place in carbon nanotubes inside those neurons, the fact that consciousness arose at the same time and resides in the same realm as language suggests something more than a casual relationship. In fact, the best explanation for the rise of human consciousness may come from the opening line of the Bible: *In the beginning was the Word.*

A COMPUTATIONAL CONNECTION

Every era has its dominant metaphors, lenses through which we look at the world. The Enlightenment used Newtonian physics, the late Victorians evolution. In the early twentieth century Einstein's Theory of Relativity and Freud's theory of the unconscious colored everything from art and literature to everyday morality. In the late twentieth century it was the Heisenberg Uncertainty principle.

In our time, the dominant metaphors come from the world of computers and networks. If, as we will see, men and women of the Enlightenment saw human thought as a ghost inside of an automaton, we imagine it as a very powerful and sophisticated computer motherboard, or as being located on a silicon chip inside a powerful microprocessor.

The brain's three main functions are logic, memory, and input/output—and when we look at the human brain through early twenty-first-century eyes that's what we see. To us, the human mind seems less a supercharged analytic engine and more a multiprocessing computer that balances these three functions of *logic* (the processing of data to produce new ideas and understandings), *memory* (with which we define ourselves and organize vast realms of experience), and *input/output* (the former combining sensory input with abstract sources such as the printed word; the latter emerging as language and art, ideas, and action).

As metaphors go, the image of the human brain as a modern multiprocessing computer isn't a bad one, and is better than any that came before. That helps explain why many people are convinced that this path will eventually lead to real "thinking machines." But for now, at least, the brain-as-computer remains an analogy, not a final explanation. A lot of features in real human brains are still missing from computers, including the ability to "heal" after a crippling physical injury, the enhancement of regularly used pathways to speed access, and an inherent sense of purpose and survival that can be found in even the simplest animals.

But most of all, there is the lack of consciousness in computers—indeed, fully expressed, in anything but us. Without consciousness, modern man becomes nothing more than a super-capable, and shockingly vulnerable, *Homo ergaster.* In other words, he likely wouldn't have been born; and if he had, he wouldn't have lasted for long, much less rule the natural world. It is consciousness that empowers those higher brain functions to reach the fulfillment of their potential.

To be conscious and self-reflexive—to know that you exist as an independent thinking being—is to also recognize that you are just a tiny, and very fragile, speck in a very large and dangerous world. It is to also know that you are mortal, and that one day you will die and all that is in your head will die with you. And in that collision between a knowledge of death and a sense of a purposeful, valuable self, there also emerges a sense that the universe has a larger purpose in which you may be a tiny part, but to which you can still make a contribution even after your death. You want to convey what you've learned, the wisdom you've hard-acquired, to those who will follow and honor your memory.

With his larger cerebral cortex, archaic modern man had the capacity to develop complex languages. But it was the increasingly complex demands of daily life, combined with consciousness and this irresistible call to meaning and purpose, that drove language forward toward realization.

We learned to talk because we now had things to learn . . . and stories to tell.

Still, at the beginning, those stories were by necessity pretty simple. Research into animal language over the last few decades has shown that the verbal communications of many animals—from birds to dolphins and whales to monkeys—can be remarkably sophisticated. So what used to look like a singular achievement by mankind now appears more like just the latest advance in a long continuum stretching back to the chirping of crickets. And there is some truth to that: Neanderthal Man, and certainly *H. ergaster*, probably spoke in a manner that wasn't much more sophisticated in design than, say, a humpback whale, and in content was little more than modern gorillas trained in sign language can produce.

But at the same time, it would be a mistake not to recognize that a profound discontinuity with the past takes place the moment modern man opens his mouth. That crumpled neural sheet of the cerebral cortex, with its vast memory capabilities, powerful facility with language, and, most of all, consciousness with its will to power and immortality, could do things that no other brain ever could. It could organize thought using logic learned from causal relationships observed in the natural world; it could synthesize new ideas through the metaphorical linking of two diverse notions, and it could imagine, perfect, and test scenarios about the world in order to more accurately understand reality.

No other creature had ever accomplished this; no creature, including earlier hominids (save perhaps Neanderthal Man), had ever known the concept of "I" or speculated into the future or used logic (deductive to understand the present, inductive to predict the future) as a path to truth. Indeed, no other animal had ever formulated the very notion of Truth. And as the allegory of the Garden of Eden suggests, to know truth is to also know falsehood. For good (the ability to create fictional realities) and evil (the destructiveness of an alienation from one's self and others), modern man now knew how to *lie*.

With a real language (also for the first time in the history of life on Earth) complex memories both old and new ("new" memories being ideas) could be transmitted from one member of a species to another—a process made ever more facile as language itself evolved, and still evolves today, to perfect this function. Shared memories not only had a longer life span, but they could actually grow in size and utility as each sharer added to their content and then shared them again. This "common memory"—shared *wisdom*, as opposed to rote, shared skills—was almost always an improvement over what any individual had in his or her head. It meant that there was nearly always an intrinsic and obvious advantage to *talking* with one another, to sharing experiences.

Sharing, in turn, rewarded ever-larger groups of people, not only because language was easy to scale but because it also exhibited what we know today as Metcalfe's Law: *The value of a network grows much faster with the number of participants in that network.* Add another member to the tribe and you don't just gain that person's memories but also the value of their interaction with every other member of the tribe. Obviously, the infrastructure, technology, and the social harmony of a society sets an upper limit to the size of the population—eventually adding new members will create unrest or a drain on resources that will give their addition a negative value—but that only creates an incentive for the tribe, village, city, or eventually, nation to put its collective head together and come up with solutions such as better food production, law enforcement, housing, and education.

So, let's follow the string: The rise of a cerebral cortex, which enabled the genetic disposition in hominids toward language and a large memory to be fully expressed, led to the creation (somehow) of a phenomenon unique to modern man, (consciousness). Conscious men and women, anxious to serve their selfhood in the face of their new knowledge of the natural world and of death, found a competitive advantage in expanding language to become more encompassing, adaptive, and abstract. Then, in the act of using language to share memories (skills, experiences, ideas, stories) with others, modern man discovered the value of increasingly large social structures—and the advantage and power of social stratification and specialization. Finally, this specialization freed certain individuals in that society—shamans, priests, and eventually academics—to focus upon expanding both those common stores of memory and the language needed to manage them.

Whew. What is miraculous about this process is not just that it happened, but how quickly it occurred. As noted, *Homo sapiens* first appeared 200,000 years ago. Just 130,000 years later—a blink in evolutionary time—archaic modern man had already begun his migration out of Africa . . . and despite facing near extinction, still managed to explore and inhabit Eurasia and Oceania in just the next 30,000 years. Just 25,000 years after that, early modern men and women had found their way across the Bering Strait and had inhabited the Americas. Just 4,000 years after that, *H. sapiens* had begun to settle down from a hunter-gatherer existence into cities and was embarking on the first Agricultural Revolution. And 10,000 years after *that*, he left the Earth and walked on the moon. In the same historic interval, most mammals, birds, reptiles, amphibians, fish, and insects changed nary a bit.

MODERN MUSINGS

Fifty thousand years ago, about the time ancient modern man left Africa, the intellectual traits that had distinguished him from his predecessors had reached their full development. These human beings of the Upper Paleolithic were, by this point, truly "modern" humans, as their new name, *Homo sapiens sapiens* (essentially, "wisest wise man"), suggests. They were now almost indistinguishable from the people you meet today. These new men and women were capable of abstract thought—not least of which was an ability to look inward to their own consciences and to look outward to place themselves both temporally and physically in the larger natural world. They had a language, and could laugh and tell jokes, or weep over a long-lost love or a sad story. They were increasingly capable of separating truth from falsehood, or rationality from passion. Indeed, they were (and we are) the only rational animal to have ever lived. And over the course of a lifetime, they lived up to the "sapience"—the wisdom—in their name. Like us, they were tirelessly in pursuit of explanations of how things worked, and if their imaginations couldn't find an explanation of the infinite and ineffable, their imaginations could at least encompass those concepts.

And as these modern men or women sat by the fire and looked up at the vault of the night sky with its moon and stars, they would have remembered the story told by a grandmother or the village shaman about how those distant lights found their way into the heavens. And they would wonder if they could tell that same story as well to their grandchildren . . . and they would fear that the story itself might one day be lost, because memories were such fragile and unreliable things, and words were just breaths lost on the breeze, like the smoke drifting up from the fire.

And what of their own life and all that they had experienced? Who would remember *their* story after they were buried in the ground with their sewing needles and spears?

If only there were some way to make those stories, and their own memories, as strong and enduring as the rocks themselves. . . .

2
The Cave of Illumination
Memory as Symbol

If we know where to look, we can see the birth of written language reenacted every day.

For example, watch a bored little boy or girl sitting in the dirt, in a sandbox or on the beach. Very soon, the child will grab a stick—or just use a forefinger—and begin making geometric designs in the sand or dust. Sometimes these symbols will evolve into more complex representations of landscapes or living things. That the medium is sand or dirt (in other words, granular) means the markings can be quickly erased in part or whole with the flick of a hand and modified or replaced . . . a process that may be repeated scores of times in a matter of minutes.

Meanwhile, if you were to find yourself out with a hunting party or combat patrol—the closest thing we have in modern life to ancient hunter-gatherer culture—you might very well find a hunter or soldier kneeling down and, with stick in hand, drawing a crude image of the surrounding landscape, then designating the quarry or enemy with a mark, and then showing with a series of gestural lines how each member of the group will move in concert for the maximum chance of success.

In other words, with the child, hunter, and soldier this most primitive form of drawing encompasses almost all of the most basic and elemental categories of human art: geometric, gestural and abstract, imaginative, representational, communicative, tactical, and organizational. Include the comic book in the child's back pocket and nervous cross the soldier quickly draws across his chest with his hand before heading out on the mission and you complete the list with (respectively) the narrative and sacred/ritualistic forms of art.

Unfortunately, we live in a world in which the lives of children are seen as distinct from those of adults. Neither have most of us experienced either combat or a hunt. Most people have, however, *doodled* either while waiting on the phone or sitting through a boring lecture, and a minor science has arisen to read the psychological underpinnings of those shapes and forms. But even doodles seem doomed in a world of smart phones, laptops, and text messaging under the school desk.

As is the case with spoken language, it is almost impossible to guess with any precision how long humans, and hominids, have been making images. Several physiological and mental processes do seem to be prerequisites to drawing. One is symbolic thinking—or at least the ability to perform the categorical logical equation: *For all X, Y is true.* Essentially, all animals that discriminate in their eating habits exhibit a basic form of this inference. Ethologists, the scientists who study animal behavior, have long concluded that this type of thinking is mostly instinctual, but can also be the result of imprinting, imitation, and conditioning.

A cat, having eaten a bird, knows that every other bird it hunts will be edible, too. It may get a few surprises, but the inference is strong enough to survive the exemptions—and a higher-order creature like the cat (in other words, a mammal) has the capacity to refine that inference, learning to go after fat, slow robins rather than tiny, fast hummingbirds. Conversely, the puppy that eats a wasp will likely still stay away from them as an old dog.

Experiments show that many, even most, animals with an actual brain stem can learn this inference, and in some of the most intelligent animals (crows, octopi, apes) this ability is sophisticated enough that they can not only string together many such inferences into a complex, multistage action—"pile up the blocks to get to the stick, use the stick to make a bridge to the button, push the button and get food"—but they can also teach this skill to their young and thus perpetuate that new skill-memory through the generations.

A second necessary requirement for drawing is fine motor skill. That's where the famous opposable thumb of the apes comes in. Even if a lizard had the mental capacity to draw, he would have a hard time doing it. Dolphins and elephants have huge brains, but even when they are taught to hold a writing implement in their teeth or trunk, the writing instruments they are using had to be constructed by humans.

Heidelberg Man, modern man's direct ancestor, and Neanderthal Man, his older cousin, both had these fine motor skills—in spades. Neanderthal Man may in fact have been the greatest natural hunter ever to live on Earth. And certainly he had a big enough brain. But even Neanderthal Man seems to have lacked the final factor needed to create images: a sense of "thingness"—an ontology—about the natural world. The wooly mammoth and his family unit may have been the two great antipodes of Neanderthal Man's existence—and he may have even created simple hunting rituals, birth celebrations, and death ceremonies. But there is no evidence that he felt the need to immortalize those moments; his vision of past and future were too limited.

Of course, as he did with spoken language, Neanderthal Man may have indeed created some crude images but used a medium—dirt, the surface of a leaf, an animal skin, or even his own flesh—which just wasn't durable enough to survive until now. And perhaps someday (a possible candidate was found in Spain in 2011) we will find a work of Neanderthal art . . . a scratched spiral or a series of lines or even a traced hand . . . that will make us reconsider our thinking. What improves the odds of this happening is that there is good reason to believe that, in the human story, art is older than language. That's because the simplest art—such as that done today by a baby—is the product of a simple set of physical gestures. By comparison, language requires well-evolved hearing, a dropped larynx, and very sophisticated manipulation of breath and muscles in the mouth.

In fact, the genetically "oldest" humans in the world, the San Bushmen of the Kalahari in Namibia and Botswana have shown that you can perform very complex tasks without language at all, thanks to a combination of knowledge of small-unit tactics and a collection of hand signals. With crude spears and arrows dipped in a slow-working poison, San hunters are able to stalk, strike, and then track a dying animal over the course of several days—without ever really having to speak. Indeed, tracking with a Bushman, on ground that has no rocks (like the Kalahari), is almost a silent process. It seems pretty likely that the signals they use—say, a *V* shape with two fingers to identify an oryx or a spiral drawn in the air with one finger to denote a kudu—are signs they began using long before they developed their clicking spoken language.

It isn't much of a jump from gesturing those symbols in the air to drawing them in the dust or scratching them on a rock. But the crucial factor is *will.* If you have almost no sense of a future, and can't extrapolate from what is and what will be, then you have little incentive to *memorialize*—that is, to create a vehicle with which to commit something to memory—an event by drawing a picture for yourself and to share with others.

This notion of a will to represent the world, as with the development of spoken language, once again takes us back to the question of facility: When did early man have a supple enough brain, one sufficiently capable of dealing with time and symbolic logic, and one with enough sense of self, to feel the need to represent the natural world in images? Add to this a further factor: *synecdoche*—the use of a part (for example, a single elk) to describe the whole (a migrating herd). Synechdoche is taking that natural sense of inference found in most animals and lifting it up to the next level of abstraction.

Once more we find ancient modern man inching his way across the 150,000-year span from his beginnings to his transformation to *Homo sapiens sapiens*, armed with all of the cognitive tools he needs but still trying to tune them for maximum harmony and performance. And perhaps it took so long for other reasons as well, not least that mankind may have been a bit distracted by the need to survive its various near-extinctions to do anything but scratch out a subsistence living.

AN AWAKENING

But then, once again, about 70,000 to 50,000 years ago, something happened and it all synthesized. The inchoate pieces of the human brain came together on the sheet of cerebral cortex to become what was identifiably true behavioral modernity. Mankind suddenly not only had a past but a future it could imagine . . . and ponder. And just as he looked around at the world and tried to explain it with rational thought and language, so, too, did man discover, in drawing, an awesomely powerful tool to capture that world in all of its complexity.

That faux Chinese proverb "a picture is worth a thousand words"—which is alternatively attributed to the copy in a 1920s American magazine advertisement as well as the Russian novelist Turgenev—has become a commonly accepted truth because it captures something essential about images: They can provoke almost endless description and discussion. Thus they are especially ripe for symbolism—as 2,500 subsequent generations of artists, both major and minor, have discovered.

So, although in theory you can use drawing to simply *render* what you see, what results has little intrinsic value to a tribe of hunter-gatherers (or indeed to modern man: Accurate portraits don't really appear until the Romans, and pure landscapes and still-lifes not until the Renaissance). Early modern man didn't have time for aesthetics; though he would discover visual representation's potency rather quickly.

What is most remarkable about the birth of human art, once it finally began, is how fast it developed. The earliest-known human "drawings"—that is, scratches on a rock—come from Africa and are dated to about 70,000 B.C. But a more accurate date is 40,000 years ago, which is the age of the earliest-known true works of art. These Upper Paleolithic works come in three basic forms: *petroglyphs*, which are carvings into rock; *pictographs*, which are paintings on rock (initially with another soft rock, such as red ochre); and *portable* art (such as small statues, carvings, and the like). The rock art can be further defined by whether it is found inside caves (as in Europe) or on outside rock surfaces (Australia, the Americas). As for the portable art, it can be carved out of rock or carved into a hard biological surface, such as antler or bone.

Two things are especially compelling about this earliest human art. The first is that it is almost always found in "special" places—caves, tucked-away corners of exposed monoliths (such as Ayers Rock), or interred with the bones of tribal leaders. The second is just how damn good they are as artistic achievements. The little carved "Venus" statues—such as the 40,000-year-old Venus of Hohle Fels and even more famously, the 24,000-year-old Venus of Willendorf—are astonishing works. Their physical distortions, sensuality, and sense of movement wouldn't be seen again until the twentieth century, and remain unsurpassed.

After four teenagers discovered France's Lascaux cave in 1940, the first visitors were dumbstruck—even though they didn't know the paintings were 17,000 years old—by the sheer beauty of the artwork. The cave's famous "Dun Horse" is one of the world's best-known paintings not because of its venerability but because it is so drop-dead beautiful. It is, by any measure, one of the greatest works of art. And to a lesser degree, this high quality characterizes other famous Paleolithic artworks from around the world: Egypt's cave of swimmers (from 10,000 years ago); the images of giraffes, men, and horses in the Acacus Mountains of the Sahara Desert (from 10,000 to 12,000 years ago); and the deer carved on deer antlers in the Turobong cave in South Korea (from 40,000 years ago). Many of the techniques and motifs in these works weren't rediscovered until the Renaissance and some not until the Fauvists and Cubists of the twentieth century.

The unique locations and the sheer quality of this Paleolithic art suggest something very important about the rise of imagery and early modern man: While verbal language increased in value to its users the more people used it, symbolic language by comparison conferred its power through exclusivity. That explains the exclusive locations: These images were never readily available to members of the tribe (nor are they today, as the paintings—as at Lascaux—are now vulnerable to even the breath of viewers).

Instead, their viewing was probably a very rare and mystical experience, likely available almost exclusively to the patriarch, the chief, or the new breed of leader—the shaman, whose task it was to manage and control the tribe's access to the infinite, mysterious, and ineffable that their powerful brains had made possible. The shaman, the tribe's chief priest, was the doorman at the portal to life and death, the terrifying and the comforting, the gods and the demons—a role he often enhanced through the use of newly discovered chemicals (alcohol, hallucinogenics, poisons) that altered the delicate operation of his brain.

It was to the shaman's advantage to enhance the mystical viewing experience as greatly as possible for the rest of the tribe—and he (and occasionally she) also likely believed that a more intense experience meant his own greater proximity to the gods and the hidden wisdom. Hence the cave: Otherwise, why would you paint such magnificent works of art in the *dark*? But picture our Paleolithic tribesman, perhaps intoxicated in some way and terrified, being led down into the black bowels of Earth by a man known to have special access to the spirit world, surrounded by sounds and creatures covered with skulls and skins . . . and then, in the sudden illumination of a torch, encountering the "Dun Horse" or the face of his worst nightmare.

It would have been a knee-buckling moment . . . and his awe and fear of the shaman, the godlike figure capable of creating this experience, would have been unbounded.

We know, based upon analysis of skeletons, that early man was fully capable of caring for the crippled and mentally limited among their numbers—often for years. And of course the leader of the tribe, the chief, was given special privileges (though often at the price of assuming greater risk in battle or a quick and fatal retirement). But the shaman is a unique case: For the first time in mankind's story, one member of the group is given a life of advantages by the community—likely including never having to hunt or gather food, a ready supply of concubines, and social power over everyone including the chief—in exchange for the advantage to the tribe of having a member with direct access to powers that resided only in the spirit world. The shaman would, of course, have to make some sacrifices in return—isolation, physical harm from the rituals, and possibly fatal consequences for failure—but it would have been a unique and probably very rewarding life. And the rise of the shaman was a major step not only in the stratification of human society, but also in its specialization.

MASTER OF THE MEDIUM

But there is also a second specialized figure in this tableau, one whose existence is not so obvious but can certainly be assumed. This is the *artist*. It would be absurd to think that paintings and sculptures that rightly deserve to stand beside a Titian or a Modigliani were created from some untrained native talent arising out of nowhere or even by a clever politician like the shaman. You don't carve the Venus of Willendorf on the first try, or even the fiftieth, no matter what talent you were born with. No, the artists who created these early masterpieces had their talent spotted early, spent many years apprenticing and practicing their art, and only then were allowed to work in the sacred location. And all the while these artists, like the shamans to whom they were attached, must have lived a separate (and probably privileged) life apart from the tribe. The world of social distinctions—and the varying degrees of power that accrued to them—was becoming well established.

The chief, the shaman, and the artist—the leader/warrior, the wise man/priest, and the incarnator/craftsman—would have been the first to set themselves apart from the main run of the tribe. But their distinction would have been fragile. Given the privileges their roles would have commanded (then, as now, few would have noticed the accompanying sacrifices demanded by these jobs), there would have been no shortage of pretenders to those titles.

Our literary history is filled with stories of challenges, sometimes successful, to weak, corrupt, or incompetent chiefs and kings. The same holds true of tales of false prophets being exposed and brought down. But in the daily life of the late Paleolithic Age there also must have been no shortage of examples of challengers for the prestigious job of tribal artist. As with the chief and the shaman, such competition would have had the salutary effect of forcing the existing job holders to do a better job. But for the artists it would also have been an excellent motivator to never stop improving their craft—not just in technique and subject matter but also in the *narrative* of the presentation: the drama of the viewing experience, the representation of time and action, and the overall *story* embedded within the image.

This storytelling, one of the greatest achievements of humankind, in its most rudimentary form probably pre-dated humanity. Neanderthal Man had enough intelligence to act out the re-creation of a hunt or a fight, even to put on some skins and antlers to add some verisimilitude. For archaic modern man, these performances would have become multiplayer, multiact tableaux, likely complete with a narrator or singer and experienced thespians in the lead roles.

As the most talented creative figure in the tribe/community, the artist might have been called upon to help with these storytelling performances; he certainly would have been expected to surpass them in terms of quality and impact. If any of this conjecture is true, it probably helps explain why human art progressed to such a high level of sophistication so quickly. Great art offered great job security.

So did cleverness. And it probably wasn't a great artist, but rather a clever innovator who first discovered how to use a series of related images to tell an extended narrative, or create novel turns in plot or capture action or simply capture the passage of time. And the strong response to such a clever technique would have led to its quick refinement as well—not to mention a host of imitators among the tribe's other budding storytellers-in-waiting.

At this point, we are now well past the line 50,000 years ago that separates archaic modern man from true modern man—that demarcation where all those human mental potentialities finally are realized and miracles start to occur. We *Homo sapiens sapiens* were more than capable of recognizing the congruence between spoken stories and pictures . . . and the value of building bridges between them. The creation of written language had begun.

Spoken language had given human beings a powerful tool to organize memory in their own brains and to share those memories with others. At the same time, art had given those same human beings a nearly unlimited means of memorializing the physical world and retaining the many symbolic meanings within it. With written language, mankind would now be able to combine the two, creating an extension of human memory that was portable, shareable, and nearly infinite in scale and scope.

Though it happened quickly, recent evidence suggests that written language was in fact a long time coming. In 2009, Canadian scientists, studying nearly 150 cave markings in Europe, were astonished to find twenty-six symbols that had shown remarkable consistency in appearance over the course of more than 20,000 years.[1] Clearly, early modern men had found these symbols useful, had ingrained them into the texture of their societies over the course of twenty millennia . . . and, with that unique characteristic of modern human beings—that of thinking about the future—had used these characters to convey that future to their children and grandchildren.

These twenty-six symbols—it's only a coincidence, but a wonderful one, that this is the same number of letters as found today in the Latin and English languages—in no way constituted a true written language. There were no semantic rules for stringing them together, no obvious punctuation, and no clear way to bolt them together to create more complex constructions. Instead, together they might be described as a *proto-language*. But all the pieces were there to make big things happen. All it needed was the spark of necessity.

WRITING HOME

That spark came 11,000 years ago. That same powerful brain of newly modern man that enabled him to speak and draw and write also enabled him to abandon the hard life of the hunter-gatherer, settle into ever-larger social aggregations like towns and cities, and embark on the Agricultural Revolution. This, the first great transformation in the history of modern man, began in places such as Jericho. Much has been made, and rightly so, of the fact that Jericho and the other cities of the Levant and Mesopotamia were founded right at the beginning of the Holocene era. This was a period of global warming after the last Ice Age (which continues to this day) that not only produced longer growing seasons but also new strains of hybridized grain.

But less celebrated is the fact that true written language also emerged at about the same time—and this, too, is not a coincidence. Civilization and written language likely operated in a symbiotic way (as it does today), with each improvement in the speed, range, and "bandwidth" of communications making possible an equivalent leap in the size and complexity of human social constructions.

It goes without saying that it is very difficult to run a large city—a city like Jericho or, later, the nearby cities of Byblos and Sidon might have contained several thousand people—especially as the community begins to evolve toward a greater division of labor, social hierarchy, and a larger dependence upon trade. In this bustling world, important events and transactions start occurring at an ever-faster pace—far faster than one can scratch and paint drawings on a rock or a stone wall to memorialize them.

It was commerce and trade, archaeologists now believe, that forced the creation of true languages, not to mention counting and arithmetic. Trade, because it deals with inventories, shipment quantities, prices, taxes and tariffs, and payables and receivables, by its very nature produces vast amounts of facts and figures—with its importance matched only by its transience. Indeed, so much data was generated that managing it likely produced the very first regular occasions in which the mountain of information to be memorized exceeded the ability of all but a few human brains to hold it.

Until this point, written language had been mostly used for "timeless" information, such as recording important events, capturing history, and disseminating decrees from chiefs, kings, and high priests. These events can be categorized as *top-down* communication, something that occurs in every large organization today: The memos and e-mails come down from the corporate offices, but there isn't much chance of successfully sending them the other way.

The written language of commerce and trade changed all of that—and quickly—not least because money was involved. For the first time, language was being used for the true two-way communication of ephemeral information over long distances. The message may have been as simple as "I just sent you 200 amphora of olive oil at the agreed price. Please confirm receipt of the entire order and include the gold with your reply."

Pretty basic stuff, but earthshaking in its implications. More efficient commerce and over-the-horizon trade meant that these new cities could grow both in size and population—driving even more specialization, division of labor, and record-keeping. And that, in turn, created an even greater need for both literacy and skilled writers—that is, clerks and scribes. And eventually, this meant that cities had to implement some sort of educational apparatus—tutors, schools, and academies—to fill the growing demand for these trained writers.

As with storytellers before them, these early scribes enjoyed their own privileged position in society. This made pursuing such a career (and all of the education required to get there) the goal of many of the best and brightest of each generation of young men in the city.

One of the interesting ironies of the story of written language is that while it represented one of the most important turning points in the story of mankind—a fact we memorialize by dividing history itself between "written history" and "prehistoric"—it was in many ways a reactionary revolution. The reason is that spoken language, untethered to writing, was a very fluid medium—as variable as its speakers and constantly morphing into new sounds and phrasings. But the advent of writing tied down these disparate voices and channeled them into a common symbology, grammar, and orthography. On the positive side, this standardization made language much easier to learn, more accurate, and more universal, but all of that came at the cost of rapid change and adaptability, as well as (for a while, at least) rhetorical innovation.

It should come as no surprise that with mankind living in relatively isolated and remote communities, the form this written language took varied from place to place. What is stunning is just how *very* different these forms were—and how they affected the way the speakers and writers of the different languages thought, organized their brains, and viewed the natural world.

Which of these ancient communities actually came up with the very first system of written language is the subject of much debate, not least because there aren't typically enough artifacts (and most of them are tantalizingly incomplete fragments) to be able to tell whether we are merely looking at a collection of images, proto-writing, or actual written language itself.

This is certainly the case with the Jiahu symbols, carved on tortoiseshells and dating back nearly 9,000 years, discovered in Henan, China. There are sixteen known Jiahu markings, and while they do seem to be symbolic representations, there is no obvious semantic relationship among them. This suggests that the Jiahu symbols are more likely a sophisticated form of proto-writing. However, their shape and form hint so strongly at modern Chinese writing that the jury remains out until more conclusive artifacts are found.[2]

A second, equally compelling, candidate is the 7,000-year-old Vinča script from Hungary and Rumania. If the Jiahu symbols seem a distant echo of Chinese, the Vinča symbols oddly suggest Cyrillic or even Roman lettering. But once again, there is no evidence to date that suggests these markings are anything more than an aggregation of symbols. That is, they do not seem to *encode* a complete language.[3]

The third candidate for oldest written language is Early Indus script, from India about 5,000 years ago. Early Indus is more pictorial than Jiahu or Vinča and, like the others, it remains undeciphered. This language, which only appears in very short strands (three to seventeen characters), seems to resemble later Brahmic writing, but most researchers disagree that it's related, arguing that it appears to have Aramaic roots. One thing is certain: Early Indus is incredibly complicated, with more than four hundred known symbols.[4]

Interestingly, in light of the fact that spoken language began there first, written language came to Africa comparatively late. That continent's first great written language, Nsibidi, emerged in southeastern Nigeria about 3,000 years ago. Perhaps the most elegant and decorative of all of the early forms of proto-writing (it has literally thousands of characters), Nsibidi is also unique because it is the only one still in use; it even made its way to the Caribbean via the slave trade. It also has the most wonderful founding myth: The Igbo people believed it was originally taught to them by baboons.[5]

Meanwhile, on the other side of the globe in the Americas, the process of creating a written language was delayed by the migration to, and colonization of, the New World. It finally got under way about 3,000 years ago in Central America with the Olmec civilization of south-central Mexico (the folks who created the colossal stone head sculptures). It, too, has not yet been deciphered.

The most mysterious and exotic of all the early written languages must be the Andean (early Incan) *quipu*—strands of as many as 2,000 colorful knotted strings that somehow seem to encode complex information. The earliest of these *quipus* is 5,000 years old—as ancient as the earliest written languages of the Eastern Hemisphere and probably a proto-language as well.[6]

So different are these *quipus* from anything modern civilization thinks of as writing that it was generations after their discovery that archaeologists first began to wonder if they were anything more than decorative. What scholars now think is that, encoded into those colors and knots is at least a very sophisticated numbers and accounting system . . . and perhaps even language (or at least words used for labeling the groups of numbers). But as with the other early languages, it may be decades, even centuries, before we find the answers.

None of these may in fact be true written languages, but they certainly show evidence of being memory devices. You scratch symbols on rocks, horns, and shells because you either want to remember an important piece of information or you want to attach those symbols to something of value so that its ownership will be "remembered" even if you aren't there. And indeed, that's what many researchers believe these Jiahu, Vinča, Early Indus, Olmec, early Nsibidi, and Inca markings (and knottings) were for.

WORDS AND THOUGHTS

Historians of written language typically categorize all forms of writing based upon the handful of defining features that first appeared in the earliest forms of proto-writing.

Scholars generally define all true written languages as having an agreed-upon set of basic symbols, characters, glyphs (technically "graphemes"), an equally agreed-upon set of rules about how meaning is assigned to the symbols, alone or together; a spoken language that this written language maps onto; and some means to present the symbols in a nearly permanent way.

The written languages themselves are generally divided into three basic categories: *logographic, syllabic,* and *alphabetic.* Logographic writing consists of words that relate to symbolic images derived from the natural world. Nearly all of the earliest written languages are logographic, such as Egyptian hieroglyphics and early Sumerian cuneiform (the world's first dominant written languages), and modern Chinese. As one can imagine, there are no pure logographic languages because then you'd need a distinct symbol for everything, so these languages quickly began to combine symbols to capture complex concepts.

Syllabic writing, by comparison, uses written symbols—called "syllables"—that combine together into words. The "syllabary" (the collection of all syllables) of a written language is considered complete when it has an entry for every sound in the corresponding spoken language. Examples of syllabic writing are as diverse as Japanese kana and Arabic.

The most interesting of the syllabic languages may be Cherokee, devised—in a work of genius—by the previously illiterate Sequoyah in the early 1820s.[7] Tellingly, Sequoyah began this effort by trying to create a logographic language but soon found that a syllabary of eighty-five (originally eighty-six) characters worked far better. And in fact, even in the face of the sheer cultural weight of American English, Cherokee is still used today in parts of Oklahoma and elsewhere. What makes Sequoyah's decision interesting is that it makes a strong case for the idea that a culture's written language is defined in large part by how its citizens think, see the world, and remember experiences.

Alphabetic writing takes this process to its ultimate conclusion. It reduces the language-symbol relationship to individual sounds—phonemes—each of which has a corresponding symbol that can be mixed and matched in an almost infinite number of ways. This kind of writing has ultimately proven to be the most successful in the modern world. One reason for this is surely the sheer flexibility of phonemes, which allow for a precision and detail all but impossible in the other language forms. Imagine a design built with LEGOS compared to one built with cinder blocks.

Given the philosophical complexity involved—for example, adopting the idea of attaching symbols not to things or even words but *pieces* of words—alphabetic writing was not a monolithic achievement. Different cultures took the process to different levels of abstraction—that is, to the bricks (consonants) and mortar (vowels) of converting verbal sounds to individual alphabetic symbols. And while it may seem obvious to us today that consonants and vowels exist independently, it wasn't so apparent to our ancestors, who were just as likely to bolt the soft vowel sound onto a nearby consonant—a system called *abugida*, which is found in early Indian. Or one might assume those sounds to be implicit—a system called *abjad*, which is most famously found in Arabic.

There is a fourth, even more fine-grained, type of written language called *featural* writing, which not only breaks words down into sounds but also contains visual details on how these sounds should be voiced. It has been suggested that Korean Hangul is the only writing of this kind, though whether it strictly fulfills that classification is disputed.

The first form of written language that is generally recognized as being fully alphabetic, with distinct consonants and vowels, is Ancient Greek. Thus, Greek is structurally the precursor of the world's Latinate languages (English, French, Italian, and so on) and, to a degree, of Germanic and Cyrillic languages as well, making it the dominant written language of the modern world.

By an odd twist of fate, however, the leading alphabets of those Greek-descendant languages aren't from Old Greek at all. Rather, the shapes of our ABCs largely derive from a syllabic language, Phoenician, that itself was born out of an ancient Semitic language created by the Canaanites in the Sinai. The Phoenicians adopted this alphabet, modified it, and, being the greatest sailors and traders of the ancient world, spread it around the Mediterranean, creating everything from Roman to Modern Hebrew and even Modern Greek.

Its ready adoption by all of these cultures was likely due to the fact that, being almost perfectly phonetic (unlike, say, English, a phonetic language's words sound exactly like its letters are pronounced) it was comparatively easy to learn. A more subtle impetus for the widespread adoption of Phoenician—and one that was ultimately more far-reaching—was that by being so simple and learnable, it expanded literacy, replaced privileged "sacred" speech, and began to break down long-established social hierarchies.

It is a measure of the power and utility of written language that the span is just 5,000 years from the crude markings of the early proto-written languages to the exquisite Roman capitals carved onto Trajan's Column—still considered some of the most beautiful, balanced, and exquisite writing ever created by the hand of man. This development occurred in less than 10 percent of the time it took for human beings to learn to string words together into sentences.

Clearly, the value of written language was instantly evident to everyone who encountered it for the first time. So valuable that, by the evidence of its rapid adoption and even faster perfection, it was probably considered decisive to the fate of a community itself. Learning to write, at least for a culture as a whole, quickly became a matter of life and death.

ALLOYED STRENGTH

Why was writing so important? There are two big reasons, and both of them deal with memory.

The first is that this historically brief era—from 3,000 to 5,000 years ago—also happens to be the period when human society transitioned from the Neolithic ("Stone") Age into a very different era, characterized throughout most of the Eastern hemisphere—except, interestingly, China, where most of these historic transitions are blurred—by a brand-new ability to extract different metal ores from the earth and alloy them into useful new metals.

The most important of these alloys gave this new era its name: the Bronze Age.

The fact that this leap in human civilization occurred at the same time as the arrival of true written language is, once again, no coincidence. With the ability to write, human and artificial memory separated for the first time and could pursue their different paths. Memory making had begun as far back as the shamans with their cave paintings, but it could now be engaged in by anyone who was literate—and it affected everyone who was not.

Again, thanks to the Agricultural Revolution and the rise of cities, society was already becoming incredibly complex, such that only the most brilliant minds could take it all in. The early proto-languages had underpinned many of these big changes by creating an enduring social "glue" for these newer and larger human aggregations.

But the Bronze Age took the explosion of information to a new level. Suddenly written languages were vital. And even as they made possible the large, multilayered "civilizations" of this new era with their priests and traders, soldiers and artisans, so, too, did these languages need to make the next step from discrete images to full narratives—that is, to real spoken and written languages.

The ability to write meant the ability to *record* information. To remove something from one's own memory and place it into a cache of synthetic memory were it could remain, largely untended, until it was needed again. This recorded memory could also be shared with others with a precision never before available with human beings passing messages from one to another. Most intriguingly, this recorded memory could also be transmitted over seemingly unlimited distances in both space and time. With regard to space, this meant the ability to converse with others—distant kings, traders, and suppliers—so far away that you might never meet them in your lifetime. This meant a new kind of communication style, transactions, and most of all, trust. In particular, trusting people outside the tribe, the society, and over the horizon to fulfill their agreements. Writing lay at the heart of this new kind of social contract.

But the ability of writing to record memories also presented a second great advantage: to *preserve* the past. True history begins with the written word, as the past suddenly becomes not just a fragmented collection of myths and legends but a corpus of records, memories, and stories committed to the written word. Once again, it is not a coincidence that the first great stories of civilization—the Pentateuch of the Bible, *The Epic of Gilgamesh*, the Egyptian *Book of the Dead*, the *Mahabharata*, and *The Iliad*—all appeared during the Bronze Age. Most of these stories, no doubt, had been written, appended, and carried around in the heads of bards and storytellers for generations. But writing made it possible to finally put these narratives safely into storage, where they could be protected, shared, and sculpted to keep up with the newer and more complicated reality of Bronze Age life. It is even possible that an overwhelming desire to commit these stories to a permanent record may have helped to drive the formation of a true national language as, later, the Koran seems to have done for Arabic and *Beowulf* for Old English.

Needless to say, once the schism took place, there was no going back. No society has ever voluntarily renounced writing. On the contrary, it would seem that the ability to record information almost immediately sets off an explosion of *new* information, records, and stories that would be impossible to stuff back into human brains. For human beings, to think is to talk, and to talk is to write, and to write is to create repositories of memories outside of our own.

But as much as writing drove the invention of "outside" memory, so, too, did it transform how memory operated within our own brains. To understand how this worked, consider an analogy to that alchemic substance that defined the age.

To our modern eyes, the leap from working with rock and naturally occurring metals like copper to that of a man-made creation like bronze seems a simple one. But it was, in fact, one of the great turning points in human history. Copper and tin, being bright and malleable, can make for beautiful and sacred objects, but they are too soft to be of constructive use. But combine the two at about 1,000 degrees Celsius and the resulting metal is both hard *and* malleable. Most important, it can hold an edge, which means it can be used for making weapons.

Bronze was so valuable to early societies that they reorganized themselves around it. Beyond its obvious usefulness in fashioning everything from weapons to instruments to the earliest machines, bronze also brought a philosophical change to the communities that worked with it. Bronze, along with leather (another new product found in towns and cities), was a *plastic* medium. You could bend and pound and score it and shape it into as many forms as the human mind could imagine—and this sparked a whole new way of interpreting, engaging with, and participating in the physical world . . . be it through a polished mirror, a mask, a new kind of sword, chain, shield, or nail.

Writing had the same effect, at about the same time, on the human mind.

For most of us today, 5,000 years later, written language is so embedded into the operation of our brains that it is difficult to notice just how much our thoughts and our memories are built around printed words.

Much of what we've learned in school came to us from the printed page, and so does most of our knowledge of current events. Movie titles, brand names, operating instructions, and an uncounted number of other pieces of information enter our brain through the unmediated pathway of words. Typically this begins with "reading aloud," but by our teens nearly every one of us in the developed world has long ago abandoned verbalizing print and can just scan a text at a pace of hundreds of words per minute. For sheer bandwidth of delivered useful information, even video can't keep pace with reading.

Furthermore, in the modern world, *how* we obtain printed information is of ever less importance—not just the format it is presented in (newsprint, broadsides, computer screens), but the nature of the symbols themselves. Much of the world, even the nations of the Far East that officially adhere to their traditional logographic symbols, has developed parallel alphabets using Latin/Phoenician letters even as they teach their children to be bilingual in their native tongue and the emerging global form of English.

But 4,000 years ago, this was certainly not the case, and it is interesting to speculate how different types of writing played a distinct role in how we organized our brains, remembered, and even looked out at the world. In other words, when you organize your brain—and particularly, your memory—around writing, then the type of writing you use matters a whole bunch.

For example, alphabetic writing, with the freedom created by its tiny building blocks (phonemes) and flexible grammar, would seem to reinforce individualism, innovation, and a civic form of democracy, but it would also perpetually run the risk of collapsing into chaos. Logographic writing, because of the sheer difficulty of learning it, would seem to reward a more stratified and rigid society, with academic and scribe classes, but it might also feature a much stronger cultural aestheticism (every word being a painting) and naturalism. And the syllabic languages, because they are much easier to learn, would seem to be an advantage for mercantilism and trade and, because they would lead to higher rates of literacy, to greater cultural democracy.

Obviously, as history reminds us, writing isn't destiny. But there is enough correlation between the traits of these types of writing and the cultures that produced them to suggest this is more than a coincidence—that there is some causal link between a culture's writing style and the way it sees and remembers the world around it, how it orders its society, and what it values. And while many of these characteristics are fading in light of the global economy, the Internet, and mass communications, enough remains to add to the tensions of the modern world.

In other words, the invention of writing didn't come without cost.

On the positive side, it enabled human society to accelerate its development at breakneck speed. The five-millennia jump from a tribe of hunter-gatherers planting the first seed to a man walking on the moon is mankind's greatest miracle. And it was possible because writing enabled cultures for the first time to innovate in parallel, share results, and then build upon those accomplishments with even more parallel innovation.

Beginning with the great discontinuity of 50,000 years ago (the dawn of *Homo sapiens sapiens*), mankind has increasingly made its own history, participating in and influencing unfolding events so completely that it is hard to distinguish later on the hand of man from the hand of nature. With animals, cause and effect are usually obvious—for example, the loss of habitat leads to extinction, an Ice Age selects for longer hair, and so on.

But with humans, especially modern humans, this is not the case. And with written language we have a perfect example of this paradox: Was the Bronze Age sparked by the improved trade made possible by the rise of formal written languages? Or were all of these advances in human communication the result of the accidental discovery—caused by the human migration north into Eurasia after the retreat of the Ice Age glaciers—of how to smelt bronze and make everything from weapons to money?

The likely answer—a little of both—is even more unsatisfactory. But the indeterminacy of beginnings, middles, and endings is a reality with which modern man must perpetually live. So, too, is the fact that because of written language, human beings not only began to think differently from their ancestors, but often differently from one another. In the long term such intellectual diversity may be a powerful tool for evolutionary adaptation, but in the near term it is the cause of considerable human friction and misery. The Tower of Babel may not in fact be a parable about language, but about writing.

"We spend our years as a tale that is told" wrote the Psalmist, who recognized that words, devised as a tool, would in turn define us. Human beings for the first time could cast their words not just to those within earshot but down through history. And appropriately enough, among those first words are those of a king as he journeys to the bottom of the sea to find the plant that he believes will give him immortality.

He loses the plant, but his printed words—"O woe! What do I do now, where do I go now?"—call to us across more than forty centuries in *The Epic of Gilgamesh*, and he has thereby found in writing something as close to immortality as mankind may ever know.

3
Clay, Reeds, and Skin
Memory as Medium

Laziness and distrust, as much as ambition and confidence, are the true sources of human innovation. It is true today in a place like Silicon Valley, and it was true at least seven thousand years ago when, on the banks of the Tigris and Euphrates Rivers, a bored and suspicious farmer or carpenter found a clever shortcut—and changed the trajectory of history.

Not long after the Agricultural Revolution reached southern Mesopotamia, farmers and craftsmen began using markers to keep count of their inventories. No doubt this was in part to keep track of stored goods to pay workers and customers, but just as likely it was to keep from being cheated or robbed. These early businessmen made markers that represented individual units. And in order to make those markers as simple to produce as possible, they used the cheapest and most flexible material at hand: dirt.

Or more precisely, small tablets of dried mud. These clay tablets fulfilled their purpose exactly: They cost nothing, were quick to make, were easy to store, and were sturdy enough to remain stored. Soon these tablets were being made by the thousands and helped propel the creation of the Sumerian society that emerged in the region.

But there were also some problems with these clay tablets. The obvious one was that they were all alike: Did that pile of tablets represent baskets of wheat or sorghum? Is that pile of tablets yours or mine? And of course there was always a potential crisis when it rained. . . .

No one is more practical than a farmer, especially when it comes to protecting his hard-earned wealth. So, it wasn't long before some enterprising businessman began embossing his clay tablets with his personal mark. But that meant these personalized tablets now had their own intrinsic value; in a sense, they became a kind of low-value proto-money. So, being another kind of wealth, they, too, needed to be preserved and protected. And the best way to do that was to create a container to hold them.

This being the early Bronze Age, a metal strongbox was out of the question. No, the solution was to go back to the same material—clay—and create jars. Form them, fill them with tablets, seal the lid . . . and you have a nice, sturdy, and largely waterproof record-keeping system for your business. Safer, too, in a world full of people trying to cheat you in business.

There was only one problem: Once you closed that jar, you were back to depending on your own memory to remember how many tablets were sealed inside, which defeated the purpose of the tablets-and-jar scheme in the first place. Once again, a clever businessman came up with a solution: He could simply make some marks on the outside of the still-soft jar that was equal to the number of tablets inside.

And then came another turning point in the history of memory. Only a lazy soul would have realized that once you had the marks on the jar, *why did you need the tablets inside*? And for that matter, *why did you need the jar* if you could just make the same marks on one of the tablets?

Thus the simple tally mark on a slab of soft clay became the medium for counting . . . and for the next thirty-eight centuries, the people of this society—which, with the help of this new recording system, would become the mighty Sumerian civilization—used this technique as their universal medium of permanent record.

The original marks on these clay tablets were essentially miniature drawings, similar to the owner's mark. But drawing on clay, as any first-grader knows, is not the easiest thing to do; chunks pile up from the groove, and the medium is so thick that it snaps the stick or reed you're drawing with; it's a serious problem. And so the Sumerians, in one of the great intellectual leaps of the Bronze Age, turned away from the little images—the pictographs of logographic writing—and developed a simpler and more flexible syllabic form of writing. In the process, they cut the number of symbols in their language from more than a thousand to less than four hundred.

But even more important, in simplifying their language—not to mention dealing with the limitations of the chosen writing medium—the Sumerians also, over the course of a couple thousand years, radically simplified their writing *style* as well. Once again, as every child knows, the easiest way to make an image in mud or clay is to take a stick and poke it in a decorative pattern. And that is exactly what the Sumerians did, albeit in a high-volume, standardized way: They took their little tablet of soft clay, put it in a little tray to keep the rectangular form from being distorted, and made little marks on its surface with a thin, wedge-shaped stylus usually made from a reed.

The result, which we call *cuneiform*, was the first mass-produced written language in human history. Indeed, the clay tablets bearing this writing—turned nearly to stone by later serendipitous fires that acted like kilns—are among the most ubiquitous, and treasured, artifacts of Bronze Age civilization.

Developing cuneiform was probably a pretty straightforward process for the Sumerians, given that Mesopotamia was the farming and trading capital of the world. By five thousand years ago, businessmen and farmers were already accustomed to making hash marks on jugs and tablets to keep count of their goods. Soon, as their numbers grew large, they learned to create new symbols to represent *groups* of numbers—like the proverbial prisoner who marks every fifth day with a diagonal line across the previous four marks.

Armed with this notion of using variants on existing symbols to represent subsets, sets, and supersets, the Sumerians probably found it pretty easy to first turn their little drawings into simplified symbols of those drawings and then to use those symbols, alone or in combination, to create more complex thoughts. In fact, it took only about four hundred years, from 3000 B.C. to 2600 B.C. for the early pictograms to evolve into the highly abstract writing we associate with cuneiform today.

The result was that daily life in ancient Sumer would have looked oddly like life today, with everyday folks sitting in shops and on benches, punching the surface of their little handheld tablets. And we can assume that experienced writers back then could have written nearly as fast in the up-down, left-right syllabic style of late cuneiform as we write in cursive alphabetic Latin style today, and probably faster than in traditional Chinese. In other words, the Sumerians, for the first time in human history, had a "permanent" writing medium that was portable and cheap, and a writing system that reduced the time from thought to printed word to a matter of seconds—not the hours and days of traditional carving in stone. And all it took to erase a sentence or message was a scrape of the thumb or a splash of water.

WORDS OF FREEDOM

One of the overarching themes of this book is that the evolution of memory, both human and artificial, is also the story of the advancement of human freedom. Every great breakthrough in our ability to capture, store, and recall knowledge has brought an ever-greater expansion of freedom and social equality in its wake.

More than forty centuries after the Sumerians created true writing, Sir Francis Bacon used that craft to write one of the best known (and least understood) aphorisms: "Knowledge is power."

Most people who read that line assume that it describes a personal phenomenon: that social power and influence come from education—that is, the quantity and quality of learning that fills one's memory. Thus, the more knowledge a person has stored in his or her head, the more power that person will have over his or her career, relationships, and the quotidian activities of life. Moreover, this knowledge will also enable that person to better deal with novel, unexpected, and even dangerous situations—for example, the Eagle Scout has a better chance of surviving being lost in the woods, the trained race driver in escaping a car crash, the sailor in getting through a storm at sea.

True enough. And Bacon himself, one of the brightest human beings who ever lived, would seem the embodiment of his own truism. But he was more than just a great essayist and philosopher of science. Francis Bacon was also the Lord Chancellor of one of the largest empires in history—and in that role, dealing with the murderous, scheming court of King James I that in the end would bring him down, Bacon was expert in wielding great political power. And so, as much as anyone who ever lived, he understood that *real* power—the kind that rules nations, changes the lives of millions, and turns the trajectory of history—was also the product of knowledge. Power was the payoff for having one's brain packed with rules and laws and secrets and a deep understanding of the lessons of history.

It follows, then, that any new art, craft, tool, or technology that enhances the ability of human beings to accumulate knowledge—either in their own memories or in some other easily accessible place—will result in a greater acquisition of power by those who take advantage of this breakthrough. And if more people have a chance to use a breakthrough—thanks to lower costs of production or greater ease of use—the accompanying power will also be more widely disseminated throughout society . . . resulting in greater equality, freedom, and ultimately, democracy.

Writing on clay tablets in cuneiform was just such a history-changing breakthrough. The birth of written language 45,000 years earlier had begun to divide society into those with secret/sacred words and images—the shamans, priests, and rulers—and those without. Because carving symbols into rock was a long and laborious task requiring considerable skill, writing—and the knowledge it encoded—was reserved for the powerful and privileged few, used mostly (as with cave tableaux) to strike awe into observers and further consolidate their own power.

But in ancient Mesopotamia, clay was everywhere, as were reeds. And the cuneiform syllabic language, especially in its early pictograms, was relatively easy to learn. With this combination of availability of technology and ease of use, it was impossible for the ruling class to maintain their monopoly on knowledge—and thus on absolute power. This, of course, didn't stop them from trying, and Sumer was hardly a modern democracy . . . but the evidence of the thousands of clay tablets that have miraculously survived to this day is ample evidence that, to borrow from a famous Mesopotamian story, the genie was out of the bottle. Writing became not only a popular tool but a highly competitive profession. As an old Sumerian dictum says: "He who would excel in the school of the scribes must rise with the dawn."

Genies work magic, and the real magic of the cuneiform and clay revolution is that it unleashed human imagination more than ever before. The zenith of that creative explosion is a collection of twelve clay tablets, from the library of the seventh-century B.C. Assyrian (the successors of the Sumerians) king Ashurbanipal. These tablets contain the world's oldest story, *The Epic of Gilgamesh.*

OLDEST ACQUAINTANCE

Only a society that has cast its language net wide enough to capture that rarest of literary talents could produce a story like *Gilgamesh.* And it takes a high level of individual freedom to allow such a story to be told—one that portrays not only the expected greatness of its protagonist, King Gilgamesh, but also his cruelties, his mistakes, and his not-insignificant flaws. This is true cultural confidence, and it explains not only the story's profound impact when its component poems first appeared 4,500 years ago but also why Gilgamesh's journey still speaks to us so deeply today. It is his agonized words about the loss of immortality that concluded the last chapter.

Gilgamesh's most obvious influence was to set the form and standard for a new kind of storytelling that swept the ancient world. It's echoes, found in later great historic poems like those of Homer,[1] induced other, later cultures to devise their own national epics and then—once their own written languages had reached a comparable level of sophistication—to write down those stories. It is probably no coincidence that the Great Flood story told to Gilgamesh by the immortal Utnapishtim and his wife—the first written record of this event—is echoed in the Bible by the story of Noah and his wife. By the same token, Gilgamesh's journey, the many creatures he encounters, and the enchantress Ishtar, find a second life a thousand years later in Homer's *Odyssey.* If, as is said, all fiction is the story of a journey, then *The Epic of Gilgamesh* built the mold, which explains why elements and themes from the poem constantly reappear in novels, films, plays, and television dramas even today.

But *Gilgamesh* also had a second, more subtle but equally powerful effect on the story of mankind. As we will see, it forever tied the present to the past; launching humanity on the even larger epic we call history.

Gilgamesh is a great work of art because it combines the deeply strange with the comfortably familiar. Some 250 generations later, the story still has the capacity to entertain, thrill, surprise, and even terrify readers.[2]

Parts of the story are surpassingly strange, seeming to breathe the dusty old air of a long-gone world: giant talking scorpion-men, a magical cedar forest that induces visions of the future, a homicidal Bull of Heaven . . . There are also moments in the narrative when the characters (and, for that matter, the author) make inexplicable decisions based upon an ethical framework we no longer share—a reminder that the ancient world was often far more alien than we imagine. And there are moments of sheer terror: Gilgamesh's seven-league (twenty-one-mile) walk through a perfectly dark void, never knowing what the next step might bring, is, in its evocative but offhand way, as scary as anything written since.

The anonymous author(s) of *Gilgamesh* did something never before accomplished, and something rarely achieved since: He captured his world in a kind of dramatic shorthand . . . and then flung that memory thousands of years into the future. It works, and will probably always work, because if Gilgamesh's actions and motivations can sometimes seem opaque, at other moments he can be shockingly familiar. He is self-obsessed, casually cruel and manipulative, noble, brave, and afraid to die—and thus stands as a perpetual reminder that even though the cultural details change, human nature rarely does.

Gilgamesh is the oldest person we "know." He stands at the beginning of history, pointing backward at the shadowy world that existed before writing bound together all of humanity, living and dead.

Unlike countless later heroic epics, Gilgamesh doesn't start out either noble or heroic; in fact, he is basically a monster and a tyrant. And the story itself doesn't begin with his birth, but instead, like *The Iliad* and the Gospel according to John, starts off in the middle of the action. And like later characters from Hercules to Tom Jones to Augie March, *Gilgamesh* is a kind of bildungsroman, in which the young hero, forced to endure a series of potentially fatal tests, emerges on the far side older, wiser, and in almost every way transformed.

At the beginning of the epic—at least, from what we think is the beginning based upon the surviving tablets that were discovered by Hormuzd Rassam in 1853 in the remains of King Ashurbanipal's library—Gilgamesh, the king of Uruk, is the familiar case of a young man drunk with power and abusing his title. Two-thirds god and one-third man—the admixture symbolizing both how far humanity has come and how far it has to go—Gilgamesh is a terror. When he isn't with temple whores, he is deflowering local virgins on their wedding night. As for the young men in his kingdom, he diverts them from their duties and wastes their energies on playing games where he can show his prowess, on having them do meaningless tasks and, apparently, making them near-slaves on endless public works projects.

Needless to say, the Sumerian people have grown sick of Gilgamesh and, knowing they can't defeat and topple him, decide to distract him instead. To do this, they make a plea to their gods to devise an equal and countervailing force to their king. The gods respond by essentially cloning Gilgamesh, creating a hairy, savage man named Enkidu who roams the forests raiding hunters' traps and who embodies the dark, atavistic side of the king.

Reading *Gilgamesh* today, you don't have to be an amateur psychologist to appreciate that Gilgamesh and Enkidu are two sides of the same man—one modern and the other primitive—and that the arrival of the id-like Enkidu liberates Gilgamesh to find his noble self and emulate the gods. Trained as we are in movie plots, it also seems obvious that Gilgamesh must destroy his "wild man" to move on.

But the ancients were far wiser than this. The Sumerians lived every day surrounded by wild men and knew they could never be fully defeated by the sword. Thus, the plot takes an unexpected turn: The long-suffering trapper hires a temple prostitute to seduce Enkidu and bring him to the city. The trapper then takes Enkidu to a shepherd's camp where he is given human food and turned into the camp's night watchman. Sex, food, and military duty—it is a strategy for dealing with barbarians that the Romans would have appreciated.

Particularly unexpected is how the two half-men finally meet. When Enkidu hears that King Gilgamesh is about to rape yet another new bride, he rushes into the city to stop him. He meets the king in the doorway to the wedding chamber and the two men engage in a vicious fight—the voice of the Old World reminding the representative of the New Order that not all of the rules can be rewritten, that sacred acts still must be preserved.

Here things start to get strange again. Gilgamesh unexpectedly wins the fight; Enkidu acknowledges this, and the two men become best friends forever. No doubt to the community's relief, Gilgamesh suggests that the two of them head off on a big adventure—in particular, to kill Humbaba, the monster ogre of the Cedar Forest. This is a little more than the local leaders had in mind and, faced with the prospect of the power vacuum created by a dead king, they try to talk the two men out of this suicidal quest. But the elders are ignored.

In the end, after ignoring or misreading numerous omens and dreams warning them off, the two men encounter the horrifying Humbaba and, with the help of the gods, capture him. When Gilgamesh—showing either the conscience or the weakness of modern man—takes pity on the monster, Enkidu forces him to kill Humbaba. Then, in an appalling act of heresy, they take the ogre's head back to Uruk on a raft they've crafted out of the sacred cedars.

As his act of near mercy shows, Gilgamesh has begun to change. And when, after being welcomed as a hero, the king is approached by the goddess Ishtar (playing the role of Odysseus's Circe calling him back to a more violent and self-indulgent life) he spurns her advances. Furious, Ishtar calls on her father, the god Anu, to send the "Bull of Heaven" down to Uruk and avenge her humiliation. He does just that, and the city is nearly destroyed. In the end, Gilgamesh and Enkidu slay the bull, becoming even greater heroes to the citizens of Uruk.

But even though most of the gods have supported the two men in their adventures, this second slaughter of a partly divine creature cannot be endured without an equivalent death—or symbolically, a sacrifice to the old world that is about to disappear. The sacrifice will, of course, be Enkidu, who ultimately must die with his world. Beset with a series of increasingly horrifying nightmares that present an image of the afterlife that is among the most disturbing in world literature, Enkidu—lamenting that he has to die in bed like a modern man rather than in a battle like a true warrior—withers and dies.

A shattered Gilgamesh is inconsolable. He has, after all, been forever severed from his older, more natural, and perhaps "truer" self. He raises monuments to his friend and wanders the forest looking for Enkidu's spirit. But as much as his grief is for the loss of Enkidu, it is also partly due to his painful realization of his own mortality. For the first time, he has felt the cold breath of death and now fears his own end. As will later be the case with Adam and Eve, whose story will be written in another thousand years (and whose Garden of Eden is, perhaps not coincidentally, in about the same location as Uruk), the price of knowledge is a familiarity with death.

Obsessed with escaping death, Gilgamesh embarks on his greatest, solo, quest—symbolically the same quest of civilization itself: to gain immortality. Only two people in Gilgamesh's world are known to have been granted eternal life by the gods: a survivor of the Great Flood named Utnapishtim (which translates appropriately as "the far away") and his wife. Gilgamesh vows to find them. The resulting journey, with its lions, endless darkness, scorpion-men, even a Charon-like ferryman, is of immeasurable importance to the future of world literature.

But Utnapishtim proves to be an ordinary man given an extraordinary gift during an equally extraordinary time (the Flood). A shattered Gilgamesh even fails what seems a simple test (staying awake for seven days) to see if he really deserves immortality. As a dejected Gilgamesh prepares to leave for home, Utnapishtim takes pity on him and tells the king about a plant that grows on the bottom of the ocean that will restore his youth.

With a clever strategy, Gilgamesh manages to retrieve the plant. But later, when he isn't guarding it, the plant is stolen by a giant snake—which, to prove the plant's power, sheds its skin as it slithers away. Gilgamesh has failed in his quest. He sets off for Urak older and wiser. But then, as he once again sees the great walls of Uruk, Gilgamesh's spirits lift at the sight of what his people have accomplished. He must learn to take consolation in that.

The message of *The Epic of Gilgamesh* is as important now as it was then: *You will die.* But if you live your life intensely, your story—as words—will live beyond you in the memories of others. And if you live your life with greatness, those words may ring down through the ages, the words cheating death even though your body cannot.

Gilgamesh sought to live forever, and failed in his quest. But perhaps he was looking for immortality in the wrong place: *Because* of his failure, he has come as close as any mortal in history to achieving it.

As for *The Epic of Gilgamesh*, the subsequent millennia have revealed it for what it really is: a memory poem. It is a celebration of the power of language and writing, a threnody to a lost, older and more elemental world that lived almost solely in the present—and a reminder of the limitations of human ambition. After the death of Enkidu, his older self and the embodiment of pre-writing and prehistory, Gilgamesh (and the rest of mankind) is severed forever from the Old World. But he still has his memories of Enkidu, and once written down, he will always have them. And so, extraordinarily, the *Epic of Gilgamesh,* the first true written work of fiction, also works like a modern-style metafiction, as it seems to anticipate its own place in history.

It is this final resonance that may explain why *Gilgamesh* not only has survived five thousand years, but is still read for both insight and entertainment. It survived the decline of Sumer and its conquest by the Babylonians—surviving even the death of Sumerian as both a spoken language and then cuneiform as a written language. And because it was treasured and preserved, it survived until its rediscovery and translation in the nineteenth century. Today, there is little doubt *The Epic of Gilgamesh* will survive at least until humanity's next great transformation.

A KINGDOM OF STONE AND PAPER

Gilgamesh and a large pile of clay tablets are about all that remain in the modern human imagination of Sumerian civilization. But another culture, one that appears to have adopted early pictographic cuneiform and modified it for its own purposes, is so omnipresent in our daily lives that sometimes it seems to still exist on some distant corner of the planet.

Egypt.

Hieroglyphics are probably the most famous of all ancient written languages, associated even by little children with the pyramids and mummies, Cleopatra and Tutankhamun. And they seem to have been invented, like cuneiform, as a collection of symbols used to track data . . . and which, with the addition of a grammar, began to be linked together into complex thoughts. Because the oldest hieroglyphic symbols yet found appeared soon after the invention of cuneiform, most linguists assume the former in some way derived from the latter, although it is also possible that (as with many inventions, even today) it arose independently.

What we do know is that in very short order, hieroglyphics quickly evolved into one of the most aesthetically rich and evocative written languages ever created. Also among the most complex: From 3000 B.C. to 400 B.C., the number of different hieroglyphs jumped from a few hundred to more than five thousand.[3]

Clearly, Egyptian writing wasn't easy to learn. And it wasn't meant to be. Hieroglyphics is among the few forms of writing ever created that was specifically designed over time to be more recondite and harder to master. The pharaohs and their advisers seemed to have instinctively understood that literacy meant empowering the people . . . and they would have none of it. Hieroglyphic literacy in Egypt during the Pharaonic Era is believed to have never exceeded 2 percent (though one village of workers in the Valley of Kings, Set Ma'at—today known as Deir el-Medina—is believed to have a literacy rate of 40 percent)[4]. So, besides making the language ever more complicated—hieroglyphics eventually included everything from symbols for letters to sounds, words, and concepts—the pharaohs also restricted writing to just the "right" class of people or for use in special service of the government.

Still, for the ancient world, and in a civilization numbering several million people, 2 percent literacy was an impressive number. And, if learning hieroglyphics was more difficult (and its use more restricted) than cuneiform, Egyptian society more than compensated with an unprecedented array of cheap and durable writing *media*.

The Egyptians, of course, like all ancient peoples, knew how to carve and paint on stone. And, being history's most famous builders, they had no shortage of walls to write on. Less known is that they also quickly learned to write on clay tablets—although, because of the complexity of their symbols, this writing was usually reduced to imprinting labels and personal seals. Still, hundreds of these small tablets, many dating back to 3300 B.C., have been discovered.

But within a few hundred years the Egyptians also had discovered a new and literally more flexible material upon which to write. And it changed everything.

The papyrus plant is a form of sedge, a type of wetland plant that also includes the water chestnut, and most resembles marsh grass or reed.[5] The most salient characteristic of the papyrus plant is that the fibrous pith of its long (up to nine-foot-tall) stalks is both stringy and sticky. Prehistoric Nile dwellers had discovered that this pith could be pounded and twisted into very efficient rope for baskets, beds, and even boats.

Inevitably, somebody discovered that one can also slice that pith into thin strips, lay them side by side into one layer, add a second layer with the strips perpendicular to the first, and then pound the whole thing together into a single sheet. When the sheet was left to dry in the desert heat of Egypt, the result was "papyrus"—a cheap, highly portable and malleable medium with (after polishing with a pebble) the perfect texture to take a brush stroke.[6]

What made papyrus so revolutionary was not just how light and portable it was compared to clay tablets, or even its surprising durability, but the way it took marks on its surface. Stone you had to scratch or laboriously carve, clay you could only poke or impress, but on papyrus, you could *draw*. Now, for the first time, you could take the rich and sophisticated paintings long worked onto stone and dirt walls and commit them to a sheet that could be rolled up and tucked into your robe.

One result of this was, as we all know, Egypt became a place covered with hieroglyphics. Suddenly they were not only carved into the pyramids and painted onto stone walls but written onto thousands (perhaps millions) of papyrus rolls—not just for the information they captured but also, as with other types of ideographic writing, because of their intrinsic decorative beauty.

Hieroglyphics, which began as a pictographic language—that is, literal images used mostly for religious purposes (*hieroglyphics* is Greek for "sacred writing")—went through the normal evolution to become ideograms (that is, increasingly symbolic images to capture complex thoughts)—and just kept on going. By the time of the Roman occupation, hieroglyphics contained pictographs, ideograms, phonetic symbols, alphabet-type letters, and even slang. There was even a second, shorthand, version for the less literate.* This explains why, after the Egyptians largely abandoned hieroglyphic writing (about 400 B.C.) it would take centuries for the language to be deciphered. Scholars, beginning with the Arab historians Dhul-Nun al-Misri and Ibn Wahshiyya in the ninth century A.D., mistakenly assumed that hieroglyphics were strictly pictographs—and quickly hit dead ends.[7]

That's why the famous discovery of the Rosetta Stone by Napoleon's troops in 1799 was so important. It wasn't, as is generally assumed, because the big slab of granodiorite contained the same message in hieroglyphics, the Egyptian demotic Coptic script that replaced it, and ancient Greek, suddenly making reading hieroglyphics easy. If that had been the case, the translation would have taken Jean-François Champollion and others twenty minutes to accomplish, not the twenty years it eventually required. What the Rosetta Stone really showed these French translators was that hieroglyphics were a lot more complicated than they imagined.

---

* It was this second, shorthand, version of hieroglyphics that would start to break the pharaohs' stranglehold on writing. Once it appeared, literacy rates began to skyrocket.

Wrote Champollion about the discovery, "It is a complex system, writing figurative, symbolic, and phonetic all at once, in the same text, the same phrase, I would almost say in the same word."[8]

It was only by the end of the nineteenth century—just in time for the golden age of Egyptology (and helping to make it possible)—that scientists like Howard Carter and various treasure hunters could read with both speed and confidence the hieroglyphics they encountered on their excavations. Meanwhile, this translation breakthrough suddenly made the search for papyrus scrolls valuable in itself.

The result was the discovery, over the last two centuries, of thousands of such scrolls. Together they present for Ancient Egypt the most complete memory of any of mankind's founding civilizations . . . one reason why Egypt is so much more "present" to us than, say, Sumer or Mycenae. Individually, some of the scrolls that were found also make an important contribution to humanity's collective legacy. They include epic narratives such as *The Tale of Sinuhe* (basically the Bible's story of Joseph told in reverse), monologues like the *Instructions of Amenemhat* (narrated by a ghost, to burnish a dead pharaoh's legacy), and *The Tale of the Shipwrecked Sailor* (a more fanciful antecedent to *Robinson Crusoe*).

But the most famous Egyptian narrative—one found carefully written on papyrus scrolls in many tombs—is *The Book of the Dead* (or, more precisely, *The Book of Coming Forth by Day*).

Strictly speaking, *The Book of the Dead* isn't a book in almost any sense. It is rather the most famous of the Egyptian "funerary texts"—that is, constantly varying compilations of prayers and magic spells that were in use for 1,500 years. Though historically younger than *The Epic of Gilgamesh*, *The Book of the Dead* feels much older. The ultimate text of the death-obsessed Egyptian culture, it seems to breathe the dust and decay of mummies and long-buried tombs.

Yet interestingly, *The Book of the Dead*, in its way, is also the most modern of all of the ancient texts. There is evidence that wealthy Egyptians of the New Kingdom ordered their own custom papyri containing their handpicked selection of spells and narratives. At the same time, even as they had the scrolls placed in their tombs, these Egyptians would also have segments of the text painted on their tomb walls, thereby creating the first custom-designed, multimedia, multiplatform "books."

Papyrus made this kind of flexibility, rich imagery, and wide-scale use possible because of the unique nature of the material itself. Nobody was going to carry around a clay tablet more than a few inches across—it was just too bulky and heavy. It also suffered if brought into proximity to rain, humidity, or sweat. But papyrus was a very different substance. Sheets could be hammered together end-to-end before they dried in order to create strips of almost unlimited length and, if you aligned the component papyrus pith slices so that on one side they ran the entire length of the strip, it was possible to roll the strip fairly tightly without cracking or splitting it.

The result was the *scroll*, that historic medium that, along with the toga, has come to symbolize the ancient world. Interestingly, while we tend to think of the papyrus scroll as being a Grecian or, better yet, a Roman phenomenon, it is in fact much longer lived than that. Its earliest use as a writing medium may date as far back as 3000 B.C. And though papyrus was largely supplanted by the beginning of the Christian era, many readers will be surprised to learn that as late as 1022—that is, well into the Middle Ages—papyrus was still used for Papal decrees ("bulls"). In Egypt and certain parts of the Byzantine empire, papyrus was used until the twelfth century. That means that the last papyrus use may have just touched the birth of modern paper. That is an amazing run of near-universal dominance by one technology.

That said, papyrus wasn't entirely alone in the world. Other cultures, without access to that particular type of marsh sedge, found other writing media. For example, the Indus civilization learned to dry and smoke-treat palm leaves to create a kind of paper that was quickly adopted throughout East Asia. In the upper Northern Hemisphere, from Russia to North America, indigenous peoples eventually stumbled upon birch bark—an extremely pliant white surface—as their writing medium. In Mesoamerica, only a bit later than the advent of papyrus in Egypt, natives learned to take the fibrous bark from certain fig trees, soak it in a caustic solution to soften the fibers, and then pound it into sheets to create a high-quality paper. The Aztecs were history's biggest users of this *amate* paper and, at the time of peak production in the fifteenth century, were estimated to manufacture nearly 500,000 sheets per year.

But each of these alternative media had serious flaws, most of them related to durability. Palm-leaf paper was so delicate that anything written upon it soon had to be copied lest the original crumble away. As anyone knows who has ever worked with birch bark, it tends to split, crack, and rot—which is why it is almost impossible to date when this kind of writing began; the oldest-known artifacts date back only to the thirteenth century A.D. Only *amate* paper, beloved by Latin American artists, still survives. Its fundamental flaw was a political one: Production was tightly controlled by the Aztec monarchy and priesthood, so although it proved an effective medium for storing the memory of the aristocracy, it did little for the everyday folk in the way we see evidenced in Egypt and especially Greece and Rome.

So, for three millennia, the papyrus scroll triumphed, becoming mankind's dominant memory repository. And the enduring symbol of this achievement was, and is, the Library of Alexandria.

THE WORLD'S MEMORY

The Library of Alexandria—often conflated with the nearby lighthouse, perhaps because, intellectually, the former was just as much a Wonder of the World—was one of those ancient entities that, because it is so often imagined, we feel we know it. But in fact, the library is best characterized by what we *don't* know about it.

For example, we're pretty sure that the library was a creation of Ptolemaic Egypt, and built in the third century B.C., but we don't know whether the work began under Ptolemy I or II. We also generally believe that the library was accidentally destroyed by fire in 50 B.C. by Julius Caesar, who was burning his ships in the face of an oncoming Egyptian army—but our only source for this is Plutarch, who wouldn't write about it for another 150 years, and even Caesar himself doesn't mention it.[9] Neither do we know what the library looked like, although we do know it had gardens, lecture halls, meeting rooms, a cataloging room, and an acquisitions department. Legend has it that above the shelves of scrolls these words were inscribed: THE PLACE OF THE CURE OF THE SOUL. But was it a single story or multistory structure? Hellenistic Greek in design or more Egyptian?

In fact, we don't even know how many scrolls were in the Library of Alexandria—one estimate is a half-million, but there were probably at least several hundred thousand, though many may have been duplicates. Nor do we know how the scrolls were obtained; one legend has it that every ship entering the harbor at Alexandria had to turn over every scroll they had on board. And though it is often assumed that it was the world's first library, it probably was not; another story has it that Mark Antony gave Cleopatra 200,000 scrolls looted from the equally extraordinary, but much less famous, Library of Pergamum in Turkey.

What we *do* know is that the Library of Alexandria set out—probably for the first time in human history—to capture, catalog, and store mankind's collective knowledge and memory . . . and came very close to doing so. Even the little we know about what was in the library's archives is mind-boggling—all of the works of ancient Egypt, all of the writings of the Greek philosophers and dramatists, the earliest records of some of the world's great religions . . . All that was consumed in the fire is heartbreaking: Perhaps 90 percent of the important writing of the ancient world is likely lost to us forever. Who knows how different history might have been if we'd had the rest of Aristotle and Sophocles and all of the rest of the great works by which the ancients defined themselves.

Almost forgotten in the story of the Library of Alexandria was that, adjoining it, was the Musaeum, its name the source of the modern term. This is appropriate because, just as the library was the prototype of all that followed, so, too (with a few wrinkles), was the Musaeum. And if the library set out to gather mankind's memory in the form of the written word, the Musaeum attempted the same through artifacts. History records that it featured a zoo, an observatory, an anatomy hall . . . and one of the greatest lecture programs in history, featuring the likes of Archimedes and Euclid. In 2005, an archaeological team claimed to have uncovered the remains of thirteen lecture halls near the site, with room for a total of five thousand students.[10]

Readers of the history of science are often surprised when the heart of discovery seems to move out of Europe and into North Africa—not just in the case of the two aforementioned scientists, but also Eratosthenes determining the circumference of the Earth and Hipparchus inventing trigonometry. They can look to the Musaeum for an explanation.

The Library of Alexandria and the Musaeum are symbols, and perhaps the greatest examples, of a new kind of human memory. They were built in recognition of the fact that the media revolution wrought by the invention of papyrus had produced an information and memory explosion so great that vast quantities of important knowledge and new ideas risked being overlooked or lost forever. The solution was to gather all of this information into giant repositories for safekeeping.

But then, the sheer size of these repositories created their own problem: Even if you had all of the information in one place, there was now so much of it that you wouldn't be able to find a specific item in that mountain of data if you had ten lifetimes to search for it. That meant you had to *catalog* these scrolls and then store them in the walls of the library by subject matter and theme. And when the number of scrolls began to reach the hundreds of thousands, you not only had to develop navigational tools to find them in the right room, in the right bin, and even the right place in that bin but you had to begin the laborious task of *indexing* the content within those scrolls. This indexing is a task that has yet to be completed. On the contrary, from the day the indexing work at the Library of Alexandria began, the challenge has only grown greater by the year and will likely never end.

LOST IN THE REEDS

In papyrus, the Egyptians seemed to have found the perfect writing medium: cheap, light, flexible, and—at least in Egypt—highly durable.

That last feature was hugely important. Even more so than today, written materials were expected to be very durable. One reason was that the competition (carved or painted stone) was almost immortal; the other was that, in the era before printing—and especially with a complicated pictographic language like hieroglyphics—volume was low and production times were considerable. The fact that we have Egyptian hieroglyphic scrolls that are still in superb condition 5,000 years after they were created is a testament to the toughness of papyrus. For preserving human memory and leaving incantations to the gods, those *Book of the Dead* scrolls placed in sealed desert caves have proven to be at least as permanent (and these days, perhaps more preservable into the future) than the carvings and paintings in those same tombs.

But papyrus is not a perfect medium. It has several inherent flaws that, although minor in Egypt, would become devastating as papyrus technology moved out into the larger world.

One of these problems is texture. Like all "paper" created from long fibers of cellulose—including *amate* paper and, later, Chinese rice paper (actually mulberry bark)—the surface is quite rough. As any painter will tell you, that kind of rough texture is not bad if you are working with a brush, as one did with hieroglyphics; but as any calligrapher will tell you, it's lousy for fine writing with a pencil or quill. The result is a tradeoff: either spend a lot more money for papyrus that has been heavily pounded and polished (and is still pretty rough), or compromise on the density—how much information you could put on each sheet. This wasn't a big deal in the largely agrarian world of Egypt in 1500 B.C.; it was a very big deal in the bureaucratic world of Imperial Rome circa A.D. 250.

By the same token, papyrus, despite being a major leap in portability in comparison to its predecessor media, was still (at least by modern standards) comparatively stiff stuff. Lining up those sheets of dried pith made it possible to roll papyrus into scrolls, but not much else. For example, you couldn't really fold it, which meant that you were stuck with ever longer scrolls. A near-contemporary history of Ramses III, the "Great Harris Papyrus" written in about 1000 B.C., is the longest known: It is more than forty meters long.

Once again, the challenge became accessibility to search: To find an item in a scroll required the reader to look through everything in the text that came before that item. And in a scroll that's nearly half the length of a football field, that would be a lot of rolling and peering. Perhaps it's not surprising, then, that the scribes who created these monster scrolls occasionally got lost: Many such scrolls feature redundant pages, suggesting either that the writers were too tired to go back to see what they'd already done, or just slotted in key pages several times to save the reader the miserable task of rolling through the entire narrative. There is also the practical problem of it requiring two hands to keep a scroll open, which meant that scholars couldn't easily read and write at the same time—something we take for granted today.

But the biggest problem with papyrus was geographic. The plant itself grew mostly in the marshlands of hot, dry climates in North Africa, in the Levant, and in the southernmost regions of Italy. Papyrus paper preferred the same dry climate: It had a tendency to rot when it encountered humidity and cold. Unfortunately, that was just where Western civilization was moving. This explains why, in illustrations and films of ancient Greece and Rome you see, accurately, representations of priests, architects, bureaucrats, and politicians working from scrolls but you rarely see even fragments of those scrolls in modern museums. That's because, unlike those venerable Egyptian scrolls baking away in tombs in the Valley of the Kings, most European scrolls rotted away in less than a century. In societies like Periclean Greece or Augustinian Rome, with their thoughts turned toward the infinite and the immortal, that kind of impermanence in their written memory was unconscionable. Needless to say, the hunt was soon on in Europe for a new, more weather-resistant writing medium.

Ironically, what also propelled the search for a new medium was the rise of libraries. Storing papyrus scrolls was not only a preservation challenge, but the challenge of searching through hundreds or thousands of feet of scrolls to find something was becoming increasingly untenable by the year. Additionally, the supply of papyrus plants, and thus paper, was finite (the plant eventually became extinct in Egypt from overuse and had to be reintroduced) and the big libraries began sucking up all of the available inventory. Being left on the short end of the stick was not something Rome was going to accept for long.

Basic economics says that when you have a compromised technology in declining supply (with the accompanying increased price) that is incompatible with a popular new application, then you have the perfect environment for clever entrepreneurs to develop a competitive alternative.

FLESH AND INK

In the end, the answer came from Turkey and one of the old cities of Ionian Greece, and it was an adaptation of a writing medium even older than papyrus.

Human beings had been writing on animal skins, if only for decoration, since prehistory. But animal flesh is tricky. For one thing, it is much scarcer—or, at least it was before mankind settled down and started herding flocks. But even after you've killed and skinned the animal, the resulting soft and bloody flesh poses a whole bunch of problems before it can be used even for clothing, much less as a writing medium.

The first challenge is that in its raw form, flesh is meat. It rots, it turns rancid, and it attracts other animals and a whole lot of different insects. Furthermore, as it dries—especially if it's been heavily washed to get out the blood and gore—it shrinks and twists and makes every attempt to turn itself into meat jerky. And even if you manage to keep it flat, the resulting dried skin is as stiff as a board and can require months of use to soften up and become pliable.

The ancients countered all of this, over the course of thousands of years, by developing a number of techniques to turn skin into *leather*.[11] The process ultimately included several steps. The first was to scrape the fat layer off the underside of the flesh to prevent rot, and the hair off the top side to produce a smooth, tough surface. The cleaned skin was then soaked in a solution of rotting vegetation (such as bark); cod oil or other oil; brains, or alum to both preserve the flesh and break down the proteins in the collagen to make it more pliant. This process, called *tanning*, also removed nearly all of the residual hair. Tanning produced a kind of skin that was sufficiently soft and flexible to make into clothing, especially because its porous nature took readily to dyeing, or to being stamped and formed into a variety of goods. If boiled, the tanned leather would shrink and start to become animal-hide glue, which created a very hard material for body armor and shields.

But working with skin could also go in a different direction. Rather than be tanned, the skins could be soaked for days in the same alkaline solution of bark or beer or, best of all, lime (calcium oxide or hydroxide) and then immediately stretched and dried. This softer material could then be scraped even further (especially on the once-hairy side) or polished, then coated with fine-grained powder such as chalk and turned into perhaps the finest writing medium ever created: *parchment*.

Parchment is a beautiful material for writing or illustration; it is soft, incredibly sturdy and durable (many medieval books written on parchment are as fresh today as they were when written seven hundred years ago), and it takes ink superbly by actually melting slightly to hold the image.[12] It can be folded, a fact that we take lightly now but was a radical improvement at the time. It also allows for erasure—typically done by scraping off the image with a knife—a feature that could prove to be both a blessing for contemporary scribes and a curse for future historians. Many important works in the centuries to come would be erased (scraped with a knife or washed with milk and oat bran), recycled, and written over in what is called a *palimpsest*.[13] And unlike papyrus and rice paper, parchment doesn't depend upon a single, fragile species: It can be made (and has) from every kind of mammal skin, from cow and sheep and donkey to wild game or even squirrel (and, horribly, human flesh). The very best parchment, it was generally agreed, was made from calfskin. This high-collagen, translucent paper was called *vellum*.*

Parchment also has its weaknesses. Like papyrus, it, too, reacts to changes in heat and humidity. The difference is that while papyrus disintegrated in the colder and wetter climes of Europe, parchment would usually just swell and shrink slightly, becoming more or less pliant with the weather. A far bigger problem was availability; even in a meat-oriented culture like Northern Europe, it takes a lot of dead cows to fill a library with books—especially when saddle makers and armorers are competing for the leather. However, thanks to the three-hundred-year bookless interregnum—the Dark Ages—after the fall of Rome, the race between the cattle and sheep populations and the demand for parchment didn't become severe until about the twelfth century A.D. And by then, a new Chinese invention was waiting in the wings.

Parchment as a writing medium might have remained a relatively niche industry—even the early Egyptians and Babylonians had occasionally used parchment for special documents—confined to Asia Minor had not the Library of Alexandria once again changed the rules. To reiterate, whoever controls memory also holds enormous power, and the very existence of the Library of Alexandria posed a threat to every other civilization in the Mediterranean—not least the Hellenistic Greeks living in the great cities of what is now Turkey. Not surprisingly, then, the Greeks resolved to create their own mighty library—the aforementioned Library of Pergamum, in Pergamon—a kind of intellectual arms race against the Egyptians.

They largely succeeded—at least until, if the apocryphal story is true, Marc Antony looted the place as a gift to his mistress Cleopatra—but it wasn't easy. The biggest problem was that the Library of Alexandria, with its insatiable demand for new texts, had created a veritable black hole for papyrus throughout the Mediterranean. There wasn't much left for Pergamon, and what there was wasn't cheap. And shortages only got worse when Egypt stopped exporting papyrus altogether.

---

* Today's synthetic "vellum," used in architectural drawings and graphic art, is designed to imitate the look of real vellum.

That helps to explain why, according to the great Roman scientist Pliny the Elder, Pergamon's king Eumenes ordered a crash course in the development of a high-volume, good-quality alternative writing material. He found it in his own backyard: As early as the fifth century B.C., the Greek historian Herodotus had noted that the Ionians in Asia Minor commonly used animal skins instead of papyrus for their record-keeping. So, all that was needed was to streamline the parchment-making process for production in volume, and more important, to adopt this new paradigm—individual sheets instead of scrolls—as an acceptable medium for important writing . . . and the Library of Pergamum was back in action. So influential was this shift from plant material to animal skin, and so associated with this library was the result, that the word "parchment" is actually derived from the name "Pergamon."

THE WRITE IMPRESSION

With hindsight, the ultimate victory of parchment over papyrus seemed inevitable—just as papyrus had triumphed over clay. But it was in fact a close-run race. It certainly helped that papyrus had durability problems in Europe and that there were chronic shortages of the material just when the Roman bureaucracy was in ascendance, and that the papyrus plant later began to disappear from Egypt due to overharvesting.

But working against parchment was what we today call "the legacy problem." Most of the Greco-Roman world's writing was on papyrus scrolls, most of the literate people in that part of the world had grown up with scrolls—as had their ancestors back a hundred generations—and even the interior design of buildings was patterned to accommodate huge piles of scrolls. As we've seen in our own times with mainframe and personal computers, operating-systems software and entertainment hardware, it is not enough for a new technology to be a bit better than its predecessor. No, to get users to abandon a standard to which they have become accustomed, you need to be *a whole lot* better. And at least for the first century of its use, parchment lacked that edge. Something was missing, and it had to do with ease of use. Papyrus scrolls may have been a pain to use, but they were neat, compact, and held a lot of content in a structured format. By comparison, loose sheets of parchment, for all of their advantages, were a mess, and sewing parchment sheets end to end to create a scroll resulted in a heavy, bulky mass.

Clearly a new format was needed, one that took full advantage of the strengths of parchment.

As it happened, the search for that solution was already under way in both Rome and the Near East—and the two threads would soon converge in a classic example of synthesis in innovation.

In Rome, by 100 B.C., literate citizens had long been accustomed to writing and reading notes and messages on *cerae*—planks of wood coated with a layer of wax. To write, they would use a sharp wooden stylus, like a pencil without lead, to cut into the wax surface. And to erase—the reason *cerae* were popular, since papyrus can't be erased—they used a small spatula-like device to smooth the wax surface.

The *cera* was invented by the Greeks (it is even mentioned in *The Iliad*) as an improved version of the clay tablet, and by the time it reached the Romans several hundred years later the technology had been pretty well perfected. The classic Roman *cera* features two hinged boards, each of them with a wood-framed sheet of wax on its inside face. Armed with a stylus and a smoother, Romans could open their *cerae,* take notes or compose a few sentences, then close the covers—like a laptop computer—tuck them under an arm and be on their way.

The *cera* proved to be a very flexible tool—a combination of Post-it note, scrap paper for rough drafts, postcard, and clipboard. It survives in historic epics in the scenes of Caesar slamming his fist, with its signet ring, down onto an open *cera* being held by some flunky/scribe. In reality, because large-scale production made it very cheap, *cerae* were used by the thousands by Romans throughout the empire. In time, multiple-"page" *cerae* began to appear in order to hold ever-larger numbers of words.

As a result, in conjunction with papyrus scrolls, *cerae* created a nice one-two punch—as the equivalent of short- and long-term memory—for the citizens of the Greek states and then of the Roman Empire. Ephemera were impressed into wax on *cerae* and enduring works were painted onto papyrus in scrolls.

In theory, parchment could have simply replaced the more fragile and scarce papyrus in this equation at some supply-demand point where the price of the new technology dipped below that of its predecessor. But new technologies never work that way: They almost always begin as a replacement for an existing solution, but eventually find their own, more powerful applications.

That was true for parchment. By the first century A.D., growing levels of literacy, combined with the army of scribes and bureaucrats needed to run the empire, had created an explosion of writing. Not only was this situation made worse by the growing shortage of papyrus, but the increasing number of scrolls was becoming unwieldy. It was one thing to fill the growing ranks of libraries around the empire—Rome alone would eventually have twenty-eight distinct libraries—but the sheer size of the empire demanded a heretofore unknown degree of mobility for information.

One solution came from the Middle East. Despite papyrus's resistance to being folded and its native fragility, some scholars began taking papyrus sheets off their rolls and accordion-folding them into pleats. This made them much more portable, especially when a *cera*-like hard cover was attached. In fact, this was the format in which the Dead Sea "scrolls" were found. Needless to say, with this type of folding one could only write on one side of the sheet. Moreover, the "pages" themselves were likely to split with use, leading some owners to glue or sew one side of the stacked folds together to keep from losing pieces.

You can see where this is going. All the pieces were now in place to create something new and revolutionary; all that was needed was a writing medium that could be folded in halves, fourths, or eighths, and sewn together. And parchment fit the bill perfectly. Better yet, someone figured out that you could slice the folds on the unsewn side to create two-sided pages. The result was the *codex*, the precursor of the modern book. It was stunningly portable, flexible (because it could be erased, the codex could do the job of the *cera and* the scroll) and perhaps best of all, it made the search for information *nonlinear.* That is, you didn't have to look through thirty feet of scroll to get to some item at the end of the text; now you could simply turn to the last pages of the codex. Even better, you could number the pages and create an *index* to tell you on which pages important information lay.

With the codex, the corpus of human memory became more accessible than ever before. In Europe, at least, the parchment codex supplanted the papyrus scroll in the course of just a few generations. In the ruins of Pompeii and Herculaneum, from A.D. 79, archaeologists have found only the carbonized remains of scrolls. By the end of that century, the poet Martial was writing raves about the portability and compactness of the codex (important for someone who composed thousands of couplets of epigrammatic verse). And within a few years after that, the scroll bins in all of those Roman libraries were being replaced by bookshelves.

THE BOOK OF EVERYTHING

Technological change brings cultural change—and ultimately creates philosophical change. Within less than a century the existence of these new codices had transformed the way that Romans looked at the organization of knowledge and memory. This evolution in thought was best captured in the sixth century by "the last scholar of the ancient world," Saint Isidore, the archbishop of Seville. In describing the nature of the codex, he offers a glimpse not only of how the histories of books and scrolls overlap, but also how bound, indexed volumes had opened the imaginations of their users:

> *A codex is composed of many books; a book is of one scroll. It is called codex by way of metaphor from the trunks* (codex) *of trees or vines, as if it were a wooden stock, because it contains in itself a multitude of books, as it were of branches.*[14]

These words come from Isidore's *Etymologiae*—a Christian "epitome," or compendium of summaries of other works, and was itself one of the greatest early examples of this new metaphor of a tree of knowledge. By the time the archbishop sat down to write it, Rome had fallen and Europe was sliding into the Dark Ages. The Visigoths, having conquered most of the continent, had captured the monarchy of Spain, and Isidore had successfully struggled to convert them to Christianity. But even as he did so, his work was being undermined within the church by various emerging heresies that he fought to stamp out.

In the face of all of this, in one of the most heroic intellectual efforts in human history, Isidore set out in the *Etymologiae* (also known as *Orgines*) to preserve the entire memory of Western civilization. He filled twenty volumes of parchment with nearly 250 chapters of records. But it was a doomed effort, as he probably knew, making his attempt even more admirable. Europe was collapsing into chaos and nothing could stop it. As if an augury of what was to come, Isidore dedicated the work to Visigoth king Sisebut. And, with Isidore's death in 636, Spain, the last redoubt of European civilization, fell into darkness as well.

But Isidore's extraordinary effort had not been in vain. The *Etymologiae*—preserved by the emergent Muslim empire that would capture Seville as well as by the durability of the parchment book itself—would survive the centuries until Europe was at last ready for its return. And when it was, its pages, both a welcome recipe for cultural revival and a sad reminder of all that had been lost, would serve as a cornerstone for the Middle Ages.

By then, its pages of animal hide would find themselves competing with a new memory medium, this time from China. *Paper.*

4
The Bloody Statue
Memory as Metaphor

After decades of film, television, and historic fiction, we often feel that we know the ancient world.

We've learned to recognize ourselves in our ancestors to the point where we can sometimes feel that we know Odysseus or Augustus or Marcus Aurelius as well as our neighbor down the hall or on the next street over. We've even learned over time to accept, if not fully understand, some of the extremes in savagery (gladiators and crucifixions) and behavior (orgies and infanticide).

Yet the details of daily life in the ancient world can still surprise us. We're still shocked to learn that the typical house in Pompeii greeted visitors—and the family children—with an erotic mural. Or that a young Spartan's final test for manhood was to stalk and murder a slave.

One of the unexpected oddities that one would have encountered when visiting a major civic building or temple in ancient Rome, especially during an off hour, would have been the sight of famous figures—senators, orators, perhaps even Caesar himself—shuffling along like zombies and peering with great intensity at statues, pillars, pilasters, and other architectural features.

If you could have looked into the minds of these great men at such a moment, you would have been even more astonished, as they might well be mentally painting the statues with blood and gore, draping bull testicles over a god's outstretched hand, or replacing the carved words on the pediment with figures on agricultural production in Gaul.

This wasn't some form of collective madness, or even a bizarre and long-forgotten form of entertainment. Rather it was hard, disciplined memory work, using the most powerful tool of the age: what was called the "art" (that is, the discipline) of memory, *Ars Memorativa*. And what these men, among them some of the most celebrated figures in history, were doing alone in those public buildings was constructing some of the greatest feats of memorization ever known—feats that can inspire even more awe in us now than when it was an everyday achievement.

Prodigious feats of memory were hardly new, even in Republican Rome. After all, the bardic tradition of long, precise recitations was older than writing itself. The most famous example of this tradition, of course, was Homer. Whoever was the author(s) of *The Iliad* and the *Odyssey*, there's no question that generations of bards recited, embellished and polished those epic poems of thousands of lines in front of kings and their courts during the centuries before the tales were finally written down.

But the first great story of a seemingly superhuman memory is that of Simonides of Ceos in 500 B.C.[1] Simonides, a famous lyric poet (he even shows up in Plato) who actually invented some of the letters in the Greek alphabet, was invited to a banquet held by the Thessalian nobleman Scopus, to deliver a panegyric (a celebratory poem) about the latter's recent victory in battle. This was an easy gig for Simonides, as he already had a reputation for a powerful memory capable of remembering long spans of text.

But Simonides's performance only managed to infuriate his host, who apparently thought that the poet's many decorative references to the Gemini twins, Castor and Pollux, took some of the attention away from him. So, when it came time for Simonides to collect his payment for the performance, Scopus offered him only half the amount and snidely told him to collect the rest from the divine twins.

Before an argument could begin, a house servant appeared to tell Simonides that two men were waiting for him outside. Stepping out into the garden, Simonides found that nobody was there . . . and just at that moment, the entire banquet hall collapsed, killing everyone inside.

So mangled were the crushed bodies of Scopus and the other guests that they were impossible to identify, and thus be given a proper burial. In stepped Simonides. To the amazement of everyone in the rescue party, the poet identified each of the many bodies by the location at which they sat around the banquet table.

Simonides's feat of memory astonished the world—and set off a five-hundred-year race to discover his secrets and match his achievements.

One theory held, as with the story of magic, that Simonides had inherited a body of secret tricks from the Egyptians—a typical explanation of the era for anything remarkable, just as "the Orient" would be 1,500 years hence. Another theory, also pretty standard for the time, was that the Art of Memory had been invented just a generation before Simonides by the secretive Pythagoreans (presumably when they weren't also creating music and the Golden Mean).

There is no evidence to support either story. And, in fact, Simonides's skill may very well have been his own invention, forged from various tools of the bardic tradition, his own natural gifts, and some newly devised tricks. Whatever the explanation, the legend of Simonides proved so potent that for four centuries, imitators and admirers were as likely to attribute his work to mystical powers as to any replicable methodology. The first known mentions of the Art of Memory, probably inspired by Simonides, is in the Greek *Dialexis* (400 B.C.), and in some of the works of Aristotle.

MASTER TULLY

It wasn't until the first century B.C. that a student of the rhetorical arts, whose name is still synonymous with oratory, set out to systematically explain the Art of Memory. He was Marcus Tullius Cicero (he was proud of the fact that his last name meant the humble "chickpea") and he is perhaps the best-known figure of the Roman Republic—not least because of his heroic, if failed, rearguard fight to save his country from sliding into the dictatorship of the Caesars.[2] As every Victorian schoolboy would someday know—they cut their teeth in Latin learning his most famous speeches—Cicero was born in 106 B.C. in a hill town outside of Rome to a lower-order ("equestrian") aristocratic family. Cicero's father was an invalid and had largely devoted himself to learning and the life of the mind . . . and passed that hunger on to his son. And by an early age, Cicero was already earning attention in the region, and in the capital, for his prodigious intellect.

Like most aristocratic Roman young men, Cicero's tutors were almost all Greeks, and it was from them that he learned the three parts of classical education—grammar, logic, and rhetoric—the *trivium*. He more than anyone would popularize the trivium and would influence liberal arts education for the next two thousand years.[3]

In theory, the three parts of the trivium were supposed to be equal in importance. Logic was the art of dealing with the world as it was—of things. Grammar was the art of working with symbols. And rhetoric was the art of communicating thought from one mind to another. But for Cicero, especially after he embarked on an illustrious, and often courageous, career in the law and then politics, it was the last of these three that gained prominence.

Indeed, even though he began as a wide-ranging philosopher (he even visited Plato's Academy in Greece to study), Cicero in time began to see rhetoric, especially oratory, as the supreme art of humankind. Great oratory, he believed, with its ability to convince others to act in concert toward a common end, was the ultimate moral guide for any society, especially a democracy or republic. And thus it followed that a great orator needed to be hugely skilled, learned, and, most of all, himself deeply moral—or his gifts would be wasted or used to the wrong ends.

It's not surprising, then, that Cicero soon turned to the then-available memory tools and tricks to help him memorize his long speeches. And when these proved insufficient, he dug deeper . . . and rediscovered the Art of Memory. Being a good, pragmatic Roman, Cicero initially dismissed all of the mumbo jumbo surrounding memory training and set out, using Simonides's own words, to deconstruct his achievement. What he found surprised him: Simonides had discovered a way to go beyond the usual tricks of repetition and mnemonics (verbal tools, such as "ROYGBIV" to remember the colors of the visible light spectrum: Red, Orange, Yellow, Green, Blue, Indigo, Violet) to a new kind of methodology that was almost sublime in its use of brain architecture. As Cicero later wrote:

> *[Simonides] inferred that persons desiring to train this faculty (of memory) must select places and form mental images of the things they wish to remember and store those images in the places, so that the order*

> *of the places will preserve the order of the things, and the images of the things will denote the things themselves, and we shall employ the places and the images respectively as a wax writing tablet and the letters written upon it.*[4]

Like a good lawyer piecing together his case, Cicero began to appreciate that what Simonides had done was to take something new (an observation, a draft of a speech, a course of study) and, in a very systematic way, attach it to something well known. Frances Yates, the British historian whose landmark book *The Art of Memory* revived interest in the subject, focused primarily on the use of architectural subjects, such as temples and public buildings, as the grounding for these memory exercises, and so it is the image of Roman senators shuffling through empty buildings that comes most readily to the modern mind. But Cicero and his fellow Romans also used geometric patterns, imaginary structures, landscapes, and other familiar "places" onto which to attach the new memories. The key was to work from a source location that was well-known nature to the thinker.

The next step was the tricky part. Because the human mind tends to remember extreme images and emotions over the common and quotidian (what today we call the Von Restorff effect), when mentally attaching a new memory to an old location, the thinker has to use the boldest strokes possible—preferably related in some symbolic or mnemonic way to the topic to be remembered. Hence the use of blood, absurdities, obscenities, and strange symbolism in order to lock in the memory.

> *Nature herself teaches us what we should do. When we see in everyday life things that are petty, ordinary, and banal, we generally fail to remember them, because the mind is not being stirred by anything novel or marvelous. But if we see or hear something exceptionally base, dishonorable, extraordinary, great, unbelievable, or laughable, that we are likely to remember a long time. . . .*
>
> *[W]e ought, then, to set up images of a kind that can adhere longest in the memory. And we shall do so if we establish likenesses as striking as possible; if we set up images that are not many or vague, but doing something; if we assign to them exceptional beauty or singular ugliness; if we dress some of them with crowns or purple cloaks, for*

> *example, so that the likeness may be more distinct to us; or if we somehow disfigure them, as by introducing one stained with blood or soiled with mud or smeared with red paint, so that its form is more striking, or by assigning certain comic effects to our images, for that, too, will ensure our remembering them more readily. The things we easily remember when they are real we likewise remember without difficulty when they are figments, if they have been carefully delineated. But this will be essential—again and again to run over rapidly in the mind all the original backgrounds in order to refresh the images.*[5]

BALLS AND BLOOD

Those words are from the first great work on the Art of Memory, *Rhetorica ad Herennium*, written in about 95 B.C. and traditionally attributed to Cicero (although that authorship has long been questioned). Here, in a legal analogy supporting the case for Cicero's authorship, is the following description from the same work. Containing the images that began this chapter, it demonstrates how the architectural-memory technique works in practice:

> *Often we encompass the record of an entire matter by one notation, a single image. For example, the prosecutor has said that the defendant killed a man by poison, has charged that the motive for the crime was an inheritance, and declared that there are many witnesses and accessories to this act. If in order to facilitate our defense we wish to remember this first point, we shall in our first background form an image of the whole matter. We shall picture the man in question as lying ill in bed, if we know this person. If we do not know him, we shall yet take some one to be our invalid, but not a man of the lowest class, so that he may come to mind at once. And we shall place the defendant at the bedside, holding in his right hand a cup, and in his left hand tablets, and on the fourth finger a ram's testicles [in Latin* testiculi *can mean either* testes *or witnesses]. In this way we can record the man who was poisoned, the inheritance, and the witnesses.*[6]

Now the process is clearer. The orator, or other memory expert, embarks on what is a four-part process. First, he prepares the text, be it a speech he has written or an existing narrative such as an epic poem. The details of this preparation process aren't fully known, but the process—probably refined with practice—likely involved breaking up the text into discrete, self-contained pieces that the user had the experience to know would lend themselves to simple and powerful imagery. This technique is similar to what psychologists today call "chunking"—the ability of the human brain to hold only about seven items at one time in short-term memory . . . one reason why American phone numbers were set at seven digits.

With the text now broken into thought-sized pieces, the memorizer would then move to the underlying structure of the process. In modern computer terms, the orator would use a system of *addresses* that were second-nature to him and onto which could be hung the text segments once they were converted into symbols or otherwise amped up. Exactly how this worked isn't entirely clear. But some memorizers famously used architecture and decoration as the "loci" of their memorizations. Others may have worked from gardens or other natural features. Still others appear to have worked from paintings or other images that were specifically designed for the memorization process.

The crucial part of these loci was that they'd be so well known to the memorizer that they would be second nature—that is, they would be burned into his long-term memory so completely that the memorizer could close his eyes and "walk through" the loci remembering every feature as vividly as if it were real. Indeed, it is very possible that experienced orators and other memorizers eventually stopped visiting their locus sites and conducted their memory tours virtually.

The third step was then to actually, or virtually, walk through each locus, mentally attaching the narrative segments to design features in the manner described in *Rhetorica ad Herennium*. In this way, the orator might memorize a speech of several hours, or the scholar the poem of a thousand stanzas, not only in the right order but even with an accuracy down to the actual sentences. Needless to say, it would be a lot easier, after considerable practice, to do this memorizing at home, where a scroll's (or the dozens of wax tablets') contents could be read and placed in memory rather than lugging them through a nearby temple.

The fourth and final step is, in many ways, the most mysterious of all. Somehow all of this attaching words and concepts to architectural and other details within the locus, using all manner of strange and gory imagery, actually worked. Somehow too, all of this added glitter managed to stay intact, and in the right place, in the orator's mind. In the fourth step the whole process had to be reversed—that is, the orator had to stand in front of an audience, as Cicero did many times before the Roman Senate, and somehow walk through this memory structure, picking out one showy detail after another, converting its symbolism into a continuous flow of words . . . all while *simultaneously* adding tone and emotion and everything else needed to connect with and influence the audience before him. One would imagine the less experienced speaker would talk like someone lost in a dream.

Even more impressive is the possibility that a master orator skilled in the art of memory might even have been able to *compose* a speech this way—that is, mentally walking through his memory loci, composing text, and then immediately converting it to imagery and draping it over some nearby fixture in a real-time, multistage brain ballet that remains quite hard for us to fathom.

MAGIC WORDS

Even to the people who study it, the art of memory can sometimes seem impossible. Certainly it was incredibly difficult, which probably helps explain why the art of memory largely was abandoned and forgotten once paper, and then printing, became cheap. These memory arts are also difficult to duplicate because the ancient texts—and there were several, including not just *Rhetorica ad Herennium* but also books on the subject by Quintilian and, most famous of all, Cicero's own *De Oratore*—aren't entirely clear in the details of the process, as is often the case with valuable insider knowledge.

For example, while the "method of loci" is the best-known technique of the ancients—not least because it is so bizarre—evidence exists that these same orators also used a number of other mnemonic techniques, including the tried-and-true process, well known to actors and schoolchildren, of just reciting the same text aloud multiple times until it is pounded into one's memory.

One thing we do know is that such extraordinary feats of memory are not beyond the capacity of the ordinary human brain. The ability to quickly memorize huge quantities of data has cropped up throughout history in people endowed with what we today call a "photographic" memory. Once deemed rare, thanks to game shows, television, and Internet videos, this skill in full or partial form has proven to be a lot more common than we previously imagined. And it has remarkable durability: Even in his nineties, the great actor Sir John Gielgud claimed he could recall every line of every Shakespearian play in which he had ever appeared. More prosaically, every day in the Muslim world, hundreds of thousands of schoolchildren crowd madrassas to repeat over and over again the Koran until they have memorized every line.

Author Joshua Foer, writing about twenty-first-century memory competitions, describes one competitor using a "method of loci" technique to remember a run of playing cards that is right out of *Ad Herennium:*

> *I was storing the images in a memory palace I knew better than any other, the house in Washington, D.C., that I'd lived in since I was four years old . . . At the front door, I saw my friend Liz vivisecting a pig (two of hearts, two of diamonds, three of hearts). Just inside, the Incredible Hulk rode a stationary bike while a pair of oversize, loopy earrings weighed down his earlobes (three of clubs, seven of diamonds, jack of spades) . . .* [7]

Surely some of the ancient masters of the art of memory were in fact, born blessed with a photographic memory. But it is also just as likely that most of them weren't, and had to work with brains possessed of little more than average imagination, memory, and intelligence.

Roman orators certainly could have carefully written their speeches on papyrus paper and then used some kind of jig to hold and turn the resulting scroll—much in the way a teleprompter is used today. So why go to all of the effort needed to master the art of memory? For all of the same reasons people have always pursued mastery of a trade or skill: It offers pride of achievement, it teaches the discipline needed to pursue other goals, it distinguishes the master from the common run of humanity, and, quite often, it pays well.

Most of all, in the case of this practice, once again memory was power. In the ancient world, access to information was difficult, the tools were unwieldy, and the recall took forever. With the art of memory, however, vast libraries could reside in the human brain. And in the hands of a great orator like Cicero, a mastery of the memory arts, combined with an unequalled skill at oratory, could move whole empires.

THE LAST REPUBLICAN

But the story of Marcus Tullius Cicero is also a reminder of the limitations of memory. Armed with his aristocratic birth, superior Greek education, and matchless gift for oratory, the young Cicero quickly made a reputation for himself in Rome as a lawyer, a teacher of the Greek trivium, a translator, and even a philosopher. His gifts were widely recognized, and soon Cicero was climbing the ranks of Roman political leadership, typically reaching each new level—quaester, aedile, and praetor—at the earliest possible age until he was finally named consul at just forty-two. This was a remarkable achievement by any measure, but especially so for a figure from the lowest aristocratic class, and a tribute to Cicero's gift as an orator. In an era without any real form of mass communications, Cicero had still managed through his public speeches to become the equivalent of a media superstar. A future of fame and glory in the elite leadership of Rome now seemed guaranteed.

But Rome in the first century B.C. was an unpredictable and sometimes very dangerous place. In the early years of that century, the Republic had collapsed into civil war, ultimately leading one general, Lucius Cornelius Sulla, to march on Rome and be proclaimed dictator. Sulla proved to be a surprisingly enlightened ruler, and eventually even stepped down after making a number of positive changes in Rome's government. But the die was now cast. Rome was for the taking.

Eventually three men, all generals, rose to the top in pursuit of this ultimate prize. Crassus, Pompey, and Julius Caesar together formed the First Triumvirate of Roman military dictators. In 60 B.C., Julius Caesar even invited Cicero to join this troika, no doubt believing that in so doing he would gain an ally. But when Cicero refused, believing that it would damage any chance for a restoration of the Republic, he made a very powerful enemy. Then Crassus (the man who crushed the Spartacus revolt) was killed in battle, and Pompey and Julius Caesar set off a second civil war, each in pursuit of his own absolute power. Out of this struggle the Roman Empire was born. Rome would never again be a republic. Working through a front man in the Senate, Caesar soon put Cicero at risk of execution. And so, in 58 B.C., Cicero fled into exile to Thessalonica, Greece.

Once in exile, Cicero fell into a deep depression. The only thing that seemed to keep him from suicide was his regular letters to his lifelong friend Atticus, whom he'd first met as a young lawyer. These passionate letters, because of their great detail about politics and daily life, are today considered among of the great treasures of Rome's "Golden Age."

Cicero returned from exile after just a year—and was given a hero's welcome. But as he quickly learned (as Sir Francis Bacon would fifteen centuries hence), navigating through a treacherous age at the top of the political world required not just genius and talent but also an ability to overcome one's own personality flaws. In Cicero's case, this weakness was a lack of consistency—being a great lawyer, he was easily swayed by the next good argument. And so, over the next dozen years, even as his public popularity grew, Cicero managed to regularly align himself with the wrong side—not least the vainglorious Pompey—while at the same time losing most of his traditional supporters. In the end, Cicero returned to Rome a second time (he had again deserted the city when Caesar's legions approached), gained a pardon from his old enemy, and tried unsuccessfully to keep a low profile.

Though Cicero wasn't a party to the assassination of Caesar in 47 B.C., he was generally perceived as both an inspiration for the act and a supporter of its aims. It didn't help that Brutus, the most famous of the assassins, called on Cicero to restore the Republic even as he brandished his bloody knife over Caesar's body.

Perversely, as leader of the Senate, it fell upon Cicero to execute Caesar's last will, a task he shared with the young and ambitious Mark Antony, and which made the two men the most powerful figures in Rome. When Cicero accused Antony of shady deals he made himself a very dangerous enemy—especially when Antony rose, with Julius Caesar's son Octavian and patrician Marcus Aemilius Lepidus to create the Second Triumvirate, the officially sanctioned dictatorship that legally ended the Republic.

To try to stop Antony's ascent, Cicero—no doubt armed with the art of memory—embarked on a series of fourteen speeches, the *Philippicae*, that are among the most famous in the history of rhetoric. But for all of their power, Cicero's Philippics were no match for the combined might of a newly aligned Octavian (soon to be Augustus) and Mark Antony. Almost immediately Antony called for the punishment of all enemies of Rome, with Cicero at the top of his list. Octavian reportedly argued for two days for the removal of Cicero's name, but in the end acquiesced.

Cicero tried to run, and so great was his reputation as a friend of the Republic, that many people harbored him during his escape. But Antony was relentless, and within a year (December 7, 43 B.C.) he was caught trying to reach Greece and was executed. His death, as reported by Herodotus, was like an image out of one of Cicero's own memory sessions:

> *Cicero heard [his pursuers] coming and ordered his servants to set the litter [in which he was being carried] down where they were. He . . . looked steadfastly at his murderers. He was all covered in dust; his hair was long and disordered, and his face was pinched and wasted with his anxieties—so that most of those who stood by covered their faces while Herennius was killing him. His throat was cut as he stretched his neck out from the litter . . .* [8]

Reportedly, Cicero's last words before his throat was cut were to tell the soldier wielding the knife, "There is nothing proper about what you are doing, soldier, but do try to kill me properly"—a beautifully balanced line of prose (the epimonic doubling, with a twist, on "proper") one might expect from a great orator. He was sixty-three years old.

Not only was Cicero's head taken back to Rome, but on Mark Antony's orders, so were the hands that had composed the *Philippicae.* All three pieces were nailed to the great speaker's rostrum in the Roman Forum, facing the Senate building. And it is said that Antony's wife, Fulvia, repeatedly visited the site in order to pull out Cicero's tongue and stab it with a knitting needle in revenge for the Philippics and their orator.*

AN EMPIRE OF MEMORY

The death of Cicero marked the final gasp of the republican dream for Rome. But Rome, as a vast and powerful continental empire under the

---

* Cicero's son lived to see Fulvia replaced by Cleopatra and to personally announce before the Roman Senate Mark Antony's defeat in the sea battle at Actium and subsequent suicide.

Caesars, would dominate Europe, the Middle East, and Northern Africa for another five hundred years. And though the most famous practitioner of the art of memory was now gone, the discipline itself would grow and become more refined in the generations of orators and storytellers to come.

One vital reason for both the vitality and durability of Imperial Rome is that it enjoyed an unprecedented access to multiple forms of memory. First, there were two recognized categories of human memory: *natural*, which we think of as the common form of information storage in the brain (in other words, a random collection of stories, images, sounds, and so on) and *artificial,* which to the ancients meant the kind of disciplined and precisely organized memory "files" produced by practicing the Art of Memory.

Today, when the memorized content of our minds most likely consists of a few song lyrics and poems, and perhaps a few mnemonics ("*i* before *e,* except after *c*"), the notion of such a schism in memory architectures seems both alien and arbitrary. But the ancients, especially the Romans, took it very seriously—and awarded artificial memory with supreme importance. To the educated Roman, having a head full of memorized and stored knowledge was the equivalent of carrying around a large, yet portable, private and quickly accessible library—or a laptop computer loaded with reference material and manuals—inside one's skull.

To this natural and artificial memory, Romans could also add the now long-established forms of external or *synthetic* memory, including papyrus scrolls and parchment codices (long-term) and *cerae* wax tablets (short-term). This portfolio of memory tools gave Romans the kind of access to memories and knowledge that was unprecedented in human history. It gave the average Roman a level of erudition heretofore impossible. But it also enabled Roman society to easily build upon existing knowledge with new discoveries . . . and then disseminate those discoveries quickly throughout the empire.

The result is obvious in the astounding feats of Roman engineering and design that survive to this day. Only a culture that remembered the glory of Greece and yet was also armed with the latest discoveries regarding arches, concrete, and metallurgy could have built the Pantheon, the Coliseum and the Pont du Gard aqueduct. We may look upon the Roman Empire as a bloodthirsty tyranny, but Romans saw themselves (not without reason) as the most enlightened and capable people who had ever lived. And given all that had been wrong with the empire for centuries, some of its durability at least may be accredited to its preservation, management, and distribution of information—that is, its collective memory.

FALL AND DECLINE

The cause (or causes) of the fall of the Roman Empire in the fifth century is still the subject of speculation, mainly because the event was so catastrophic that many of the records of the era were lost in the process. Did the empire rot from within, a victim of its own success as generations of decadence and wealth among Roman citizens made them unwilling to defend themselves and ever more dependent for protection on the same barbarians they fought? Was it the inevitable decline, as with all dictatorships, in the quality of the Caesars who ruled the empire and the legions that served them? A weakened and diminished population due to periodic plagues? Or was it just the historic inevitability of the barbarian tribes, whose birth rate, energy, and ambition could no longer be denied? The use of lead in Roman water pipes? Imperial overreach that extended beyond the era's communications technology? The limitations of a plunder-based economy? The destruction of natural resources from one end of the Mediterranean to the other? The shift in economic power from Italy to Byzantium, thanks to the latter's greater access to China, India, and the rest of the Far East?

Probably all of them, to one degree or another. But for our purposes, a far more important question is not why Rome fell, but why it did so when it did . . . and why so completely?

The traditional date, designated by historian Edward Gibbon, for the fall of Rome is A.D. 476, when the Germanic chieftain Odoacer, himself a Roman general, marched on Italy and captured Rome.[9] He then turned on Ravenna, where he deposed the teenage emperor Romulus Augustulus on September 4, 476, and declared himself king.* As the first Germanic king of Italy and the Western Empire, his sword-backed coronation makes for a convenient end date.

* He didn't declare himself emperor so as not to anger the Byzantine emperor. Indeed, Odoacer officially considered himself a vassal of the emperor in Constantinople, so as not to invite invasion.

But there is another, very good reason for setting the fall of Rome on this date. The Imperial City and its surrounding empire had already been successfully invaded a number of times before, of course. The Visigoths, fleeing the Huns, had invaded in 378 and destroyed a Roman army in battle before accepting settlement inside the protection of the empire. And on New Year's Eve, 405, in one of the most unforgettable images of the age, a giant mob of Vandals, Suebi, and Alans exploded across the newly frozen Rhine River near what is now the city of Mainz, Germany, and swept through Gaul (France), down the Iberian peninsula (Spain and Portugal), and into North Africa.

Then, in 410, the mistreated Visigoths again rose up, and under the command of Alaric, captured and sacked Rome for the first time. The city recovered, but it would never again be the same, and the shift of power to Constantinople and the Eastern Empire accelerated.

After that, the disasters began to pile on. In the 440s, the most terrible scourge to date, the Huns under Attila attacked the Balkans and Gaul and threatened both of the empire's capitals. In 455 it was the Vandals' turn to sack Rome. They abandoned the city and moved on to capture the major Roman cities of North Africa. A crushing blow came in 461 and 468 when Rome attempted a pair of naval counterattacks against the Vandals, only to be defeated both times.

Now there was little left to save. The final straw came in 493, after the deposition of Romulus Augustulus, when the Eastern emperor, Zeno, concerned that Odoacer's kingdom was becoming too powerful, convinced the Ostrogoths to attack Italy. They defeated Odoacer but instead of pledging their fealty to Zeno, they established their own kingdom under Theodoric. The schism between the Eastern and Western Empires was now complete and permanent.

So why, in light of all of these misfortunes, choose 476? The answer is that before Odoacer's attack, Rome *still had a memory of itself.* The Imperial City and its environs may already have been attacked and sacked several times over the previous century, but the attackers had always *left.* The scrolls, the codices . . . the memories and acquired knowledge of the preceding millennium of Greco-Roman history were still intact. They could be relied upon as the empire rebuilt itself. And as long as this memory was accessible, Rome could recover from almost any insult.

There is a famous description of Mongol tribesmen, having breached the Great Wall and attacking the great cities of Imperial China, roaring down the streets grabbing chickens and cooking pots and horses . . . and riding right past valuable treasures that they were unable to recognize. The same was likely true of the Vandals, Visigoths, and especially the Huns, who probably knew enough to steal the gold plating off the doors of Roman libraries but were oblivious to the real treasures that lay inside.

The reason 476 proved to be such an important date for Rome is that this time the invaders *stayed*. The last Caesars may have been weak or degenerate shadows of their predecessors, but culturally they maintained a continuous line that led back five hundred years to Julius and Augustus. At the time, for the average Roman citizen, the powerful and decisive Odoacer may have even seemed a welcome change from the weak and vacillating child who officially ruled their lives. But this short-term gain came with appalling long-term loss. Rome was now ruled by Germanic barbarians—people who, for all their bravery, artistry, and tribal cohesion, nevertheless had no tradition of literacy and oratory, much less scholasticism. Inevitably, those traditions—including the art of memory, being no longer valued and rewarded under the new order—slowly faded away.

This may explain the apparent contradiction between the seeming near indifference by Romans to the fall itself—life for many people just seemed to go on as before, which has led some historians to argue against the notion of a "fall" at all—and the growing sense in the years that followed that a great catastrophe had occurred.

DARKNESS FALLS

Only a few of the most visionary men and women—most of them strong believers that God would never abandon them—could see what was coming and set out to preserve the memories of their time. In Spain, Isidore wrote his *Etymologiae*. In what is now Algeria, the bishop of Hippo, Augustine, having already written the *Confessions*, history's first great autobiography, next turned to writing *The City of God* to console his fellow believers in the face of what was to come. The preeminent father of the Western Church, Augustine was murdered by the Vandals in 430.

These three works, among the glories of Western civilization, survived the fall. So did Cicero's *De Oratore, Philippicae*, and other works, not least because of the awe they produced among later, less-erudite readers. But much was lost, and much of what did survive had been sequestered in other parts of the world before the fall or spirited away just after. And there, written in strange tongues, describing impossibly complicated practices and describing worlds long since forgotten, these texts soon became forgotten as well.

By the time the new barbarian rulers of Europe had become sufficiently civilized and cultured to want to take advantage of the vast wisdom of the ancients, it was already long gone. The papyrus scrolls, with their regular need of copying, had long since rotted away. So had the wax tablets. Only the parchment sheets survived—and they, too, were often abandoned as worthless in a world without writing. What texts did survive—in Africa, the Middle East, and Spain—were in the hands of enemies, and thus were as inaccessible as if they had been on the moon. Even Roman paintings, most done as murals, faded and collapsed with the walls that bore them.

Only the massive and magnificent architecture remained. But without the memories, texts, and tools that brought forth these engineering technologies they could not be duplicated or even repaired. Even the all-important recipe for Roman concrete, the miracle construction medium of the empire, was lost for the next thousand years.

And so the glory of Rome soon was reduced to such astonishing ruins from Britain to Persia that the locals who lived at the feet of these great, crumbling edifices looked upon them as having been created by a race of superhumans.

We call this era, the five hundred years following the fall of Rome, the *Dark Ages,* although historians tend to wince at the term because of its negative connotations. And there is certainly some truth to that concern, especially the older use of the term to describe the entire Middle Ages from A.D. 500 to 1500. Today, we know too much about the second half of those thousand years—the lively and creative "Late" Middle Ages of A.D. 1000 to 1500—to still consider them as having been lost in the darkness of ignorance.

But the Early Middle Ages, from A.D. 500 to 1000, are a very different story. It was Plutarch who named them the Dark Ages and, great poet that he was, he nailed something essential and resonant with his use of the term—so much so that despite several generations of historians refusing to use the phrase, it still is the term of choice used by average folks.

They may be right to do so. Western Europe after the fall of Rome remains the most terrifying example in recorded history of what happens when a society loses its memory. It has become a tired truism that it only takes a single generation to fail to transmit its culture to the next generation to bring about the end of civilization, but that is exactly what happened in the decades after A.D. 476. What is especially tragic is that all of the memories were still technically there, in the minds of a handful of aging men and in the thousands of scrolls and books moldering away in unused libraries. But those who could use this knowledge, the rulers, were largely illiterate or indifferent, and those who could understand it were powerless. Western Europe was sinking into a dark abyss of chaos, violence, and ignorance.

There was a bright moment midway through the Dark Ages: the coronation of the Frankish king Charles, "the Great"—*Charlemagne*—who ruled from 768 to 814. Charlemagne, the grandson of Charles Martel (the general who ran the Muslim invaders out of southern France at the history-turning Battle of Tours in 732), conquered most of Europe and then set out to restore the glory of Rome and the Caesars in his own Carolingian Empire.

Enlightened, and a dedicated supporter of learning and the arts, Charlemagne was a ray of light in the Dark Ages. But his goals were impossible: Too much had been lost, and even his long reign wasn't long enough to reinvent an entire civilization. It proved to be only a half-century-long interregnum, a brief arc of illumination bookended by two halves of the Dark Ages, each two centuries long. When Charlemagne died, the empire fell apart under the rule of his descendants, and the glow of enlightenment didn't return until the dawn of the new millennium.

LOOKING EAST

It goes without saying that even as Western Europe and North Africa were lost in the Dark Ages, elsewhere in the world other civilizations were surviving, even thriving. In Byzantium, the Eastern Empire, the rhythms of old Rome continued apace, albeit with an increasingly Oriental flavor. Rich, dedicated to diplomacy over conquest (it bought off the Huns, for instance), and ruled from the mighty (and mightily defended) capital of Constantinople, the Byzantine Empire looked as if it might last forever. And it nearly did: When it fell to the Ottomans in 1453, it had survived another one thousand years.

The Eastern Empire (still calling itself the "Roman Empire") enjoyed the unique advantage of being ruled by a brilliant and ruthless emperor, Justinian, just as the Western Empire collapsed. Seeing the power void in Rome, Justinian ordered his legions to recapture most of the Western Empire in Italy, southern France, and North Africa.

He largely succeeded, but in the face of endless counterattacks by the barbarian tribes in Europe and devastating plagues at home, Justinian was never able to consolidate his gains—likely missing the very last chance to salvage something of the remnants of the Western Empire and stave off Europe's Dark Ages.

What the Eastern Empire *did* save of the West's collected memory was all of those things already embedded into its culture: art, architecture, military science, key texts, and the Christian liturgy. With its immense riches—and despite a political system that gave its name (Byzantine) to the complex, devious, and scheming—the Eastern Empire produced some magnificent works of art. Perhaps the most famous are the Hagia Sophia cathedral (later a mosque, and currently a museum), completed in 537 and one of the world's greatest buildings—fittingly, the Hagia Sophia was dedicated to the Word—and the beautiful mosaics (including a portrait of Justinian) in the Basilica of San Vitale in Ravenna, Italy.

In terms of the history of memory, and thus of civilization, Justinian's greatest achievement was to convene a ten-man commission to revisit all of the surviving Roman legal codes, transcribe them, collate them, and condense them into a workable document. It took five years. The result was the *Codex Justinianus*. Ultimately, the commission, led by a court official named Tribonian, produced four texts: a compilation of Roman law (the Codex), a collection of extracts from Roman jurists (the Digest, one of the first encyclopedias), a textbook for students (the Institutes), and a list of new laws created by Justinian (the Novels).

Together, they became known as the *Corpus Juris Civilis* (*The Body of Civil Law*). It is the seminal work of jurisprudence in the Western world and the heart of most modern (including papal) law.

Still, in part because it sat within a complex nexus of languages and cultures, in part because it dismissed the past achievements of the Western Empire as symbols of failure, and not least because it seemed to manifest some of the larger cultural exhaustion of Imperial Rome, the Byzantine Empire is not known for its larger intellectual achievements. That includes writing: Byzantine literature, though extensive, is known for either being a pastiche of older styles or plain, unemotional recitations of facts. The latter can be useful, especially in Byzantine histories, but it doesn't make for fun reading.

As with the West and Charlemagne, the Eastern Empire had its own unlikely "middle" age of light and learning. This was the Macedonian (or Armenian, because that was the ethnicity of the rulers) dynasty, which began in 867 and lasted nearly two hundred years. The Macedonian dynasty not only marked the era of greatest power (if not size) for the Byzantine Empire, but it also reached such heights in culture and the arts—and ultimately in influence—that it is sometimes described as the Macedonian Renaissance.

Exquisite works of art were created during the Macedonian dynasty. But for our purposes, the single most important change that occurred during this era was that the West began, fitfully, *to regain its memory.* This occurred in two ways. First, a new interest in collecting started to grow. Suddenly, after four centuries of disinterest, classical studies and antiquarian scholarship became all the rage. Happily, given their declining state, old manuscripts were gathered, scrutinized, copied, and taught. And the more the Byzantines learned about not just ancient Greece but also the long-gone Western Empire, the more impressed they became with what they read . . . and the more anxious to learn more.

The greatest of these collectors was Photios, patriarch of Constantinople. Photios so loved old manuscripts that he sponsored searches for them, paid to have them copied, and ultimately established the greatest library of the age. It was a lucky moment for civilization.

The second important intellectual trend of the Macedonian dynasty was (not entirely independent from the new fad for collecting) an intense interest in compilation and categorization. Over the course of these two centuries, scores of major historians and chroniclers arose to produce endless volumes recording the past. As a result, the tenth century A.D., propelled by manic collector Emperor Constantine VII Porphyrogenitus, was the first Age of Encyclopedias. With an intense desire to organize and catalog not seen again until the eighteenth century, Byzantine academics set out to organize the known world and all of human memory, in the process creating vast encyclopedias of political science, epigrams, ancient writings, and physical science.

The most important of these encyclopedists was arguably also the greatest figure of his age. A monk, philosopher, teacher, writer, and historian, Michael Psellos was born at the beginning of the eleventh century and dominated its end. Psellos, like Cicero before him and Bacon after, was one of those wide-ranging maverick geniuses who manages to rewrite his world even as he reaches the highest levels of ruling it: In Psellos's case, he became prime minister to the emperor. Like the others, he, too, fell from grace at court (the reasons are obscure, as is much of his personal life other than the fact that, given the meaning of "psellos," he probably stuttered), and in his case he found sanctuary in a monastery. But not only did Psellos return to glory a few years later, he even managed to stay there: Psellos spent the rest of his life as a trusted adviser to a succession of emperors.

Like Photios before him, Psellos was an avid collector and scholar of antiquity. But unlike his predecessor, Michael Psellos was a man of great imagination and strong opinions. He didn't see the texts he collected as being of equal value, nor their contents deserving of equal reverence. On the contrary: When Psellos sat down to write his history of the Byzantine emperors, the *Chronographia*, he gave unprecedented emphasis to the characters of his subjects over the dry recitation of battles and monuments.[10]

But it was in his philosophical histories that Psellos's personality really came to the surface. Just as the Elizabethan research scientist William Gilbert accused Francis Bacon of writing science "like a Lord Chancellor," so, too, did Psellos fill his works with his own thoughtful judgments on historical figures normally considered beyond criticism. For example, he placed Plato in the supreme position above all other philosophers—including Socrates and Aristotle, the pre-Socratics, and the Roman and Byzantine philosophers—an opinion that not only put him at odds with his peers but even led Psellos to the brink of being charged with blasphemy. A skilled politician, he survived that scandal (he took an oath to the gods) and went back to work. He even had the courage, as perhaps the first such writer since Augustine, to fill his writings with autobiographical passages.

With his combination of wide-ranging genius, academic discipline, ambition, and ego, Michael Psellos can sometimes look like a man out of time—four centuries before his time. And it was precisely his singular position and personality that enabled Psellos to take the restoration of Western memory to the next level, from collecting and collating to actual *discernment* and *application*. It was an achievement as great as Isidore's was five hundred years before, but because it was much more subtle and personal, it has never been as celebrated.

THE GOLDEN LIBRARY

The memories of Western civilization weren't only in libraries and schools in Constantinople. Elsewhere, around the fringes of the former Western Empire, there were other, equally important caches of memories being carefully preserved by different groups, and for different motives.

On the northwest edge of Europe, often literally driven out onto the rocks in the Atlantic Ocean to escape the predations of barbarians and Vikings, Irish Catholic monks huddled in their cold barrows and copied, annotated, illustrated, and rewrote the sacred texts of Christianity on sheets of vellum. Of all of the saviors of civilization, theirs was the most miserable and thankless work of all. Forgotten by the rest of mankind, they labored on, generation after generation, waiting for the light to return to Europe. Their thankless sacrifice still haunts the modern world.

In Spain, Arab scholars were also poring over surviving memories of the ancient world. Arabic civilization, which began its expansion with the fall of Rome and the power vacuum left in the Middle East by the retreating legions, went into overdrive with the birth of Islam and the death of the prophet Muhammad in 632. Following Muhammad's command to conquer and convert the heretics, Muslim armies swept south into sub-Saharan Africa, west across North Africa, north into Syria (and eventually into Turkey), and east toward India. They also attacked Europe in a vast continental pincer movement . . . only to be thrown back (temporarily) in the east at the walls of Constantinople and in the west (permanently) by Charles Martel and the Franks at Tours.

This era of rapid expansion by the Umayyad caliphate was turned into a five-hundred-year "Golden Age" of consolidation and empire by its successor, the Abassid dynasty. It wasn't long before Arab scholars were following the train of the conquering Muslim armies, looking for their own intellectual plunder.

It was there for the taking. In Egypt, these scholars found the remains of and the successors to the Library of Alexandria. And in Spain, most famously in Cordoba, they found caches of scrolls and parchments that had been left by the Romans and preserved by the Byzantines. Unlike the Byzantines, the Arabs held no contempt for the long-gone Western Roman Empire, and so they were able to appreciate—two centuries before Constantinople—the treasure they had found:

> *Andalusia was, above all, famous as a land of scholars, libraries, book lovers and collectors . . . when* Gerbert *studied at* Vich *(ca. 995–999), the libraries of Moorish Spain contained close to a million manuscripts . . . in Cordoba books were more eagerly sought than beautiful concubines or jewels.*[11]

No one understood this miraculous gift more than Abd ar-Rahman III, known as Emir Al-Hakam II. He was a pretty bloodthirsty character, having helped his father with a notorious massacre and spending much of his reign fighting everybody from Franks to Vikings. But Al-Hakam II was also celebrated for his public works, and none was more famous than that of the gigantic Royal Mosque of Cordoba. The sheer size and beauty of the place alone would have made it famous, but what Al-Hakam II did with it carved out his place in history: He added a library.[12] And not just a small, personal library—Al-Hakam II built it big enough to hold all of the old volumes found in that part of Spain, as well as all of the new works he proposed to buy.

> *The city's glory was the Great Library established by* Al-Hakam II *. . . ultimately it contained 400,000 volumes . . . on the opening page of each book was written the name, date, place of birth and*

> *ancestry of the author, together with the titles of his other works. Forty-eight volumes of catalogues, incessantly amended, listed and described all titles and contained instructions on where a particular work could be found.*[13]

To give you some context, this library, which would have been about the size of the collection of a typical small liberal arts college, by the time it was installed in the Royal Mosque of Cordoba, probably held more total books than did all of the castles, private collections, and libraries in the rest of *Europe.* By comparison, the two other biggest libraries in Europe, in Avignon and the Sorbonne, didn't have more than four thousand volumes between them.[14]

Interestingly, the man Al-Hakam II appointed as director of the library, named Talid, turned around and hired as his deputy a Fatimid (North African) woman named Labna. It was her job to tour the bookstalls and markets of Cairo, Damascus, and Baghdad in search of the rarest texts to purchase. This is the first appearance of a woman in a key role in this history of memory—and Labna herself might be seen as the prototype of the intrepid Lady Librarian of modern movies and musicals. Eventually, the library would employ as many as 170 women translating copies of the Koran.

The new Arab obsession with books and learning soon intersected with a technological revolution taking place at the other end of the Muslim Empire. Far from the cares of Europe and its Dark Ages and endless wars, China had survived its own series of challenges to emerge as a confident Middle Kingdom.

FROM THE NESTS OF WASPS

Following the collapse, in 618, of the short-lived and tyrannical Sui dynasty, China embarked on its own brief golden age under the Tan dynasty. For most of the next three hundred years, China enjoyed considerable prosperity, technological innovation, and, unusual for that culture, extensive international trade. All of this would end in a bloody civil war that would throw the country into a century of chaos, but not before all of that innovation and trade would put one of the most important inventions in the story of memory into Arab hands.

As noted, the Chinese already had a form of paper made, as did Mesoamericans, out of processed bark and wood fiber. A court eunuch named Cài Lun is generally credited with having invented this primitive form of paper in A.D. 105 during the Han dynasty.[15] His crucial breakthrough was not the use of bark, but that he also tossed into the mix a veritable potpourri of other ingredients, including the "bast" (inner) bark of fibrous plants such as flax and hemp, old rags, and even fishnet. The stew was soaked in caustic solutions that broke everything down into its stringy components and then was poured through a screen and dried. Lin attributed his discovery to his careful observation of how wasps made their nests.

Interestingly, the first popular uses for this new invention ranged from wrapping paper to toilet paper to an actual writing medium. And in time, as is the case today, these different uses ultimately led to different grades and styles of paper. True writing paper was typically made with the finest materials, vegetable dyes, and even natural insecticides to protect them.

Paper offered so many advantages over its predecessors: It was inexpensive (and became even cheaper with economies of scale), soft, and pliant. It folded easily, which meant it could readily be bound into books, and by the year 300 the Chinese were doing just that. Paper also took ink beautifully, and so, by the middle of the Tan dynasty (*c.* 1040), Chinese publishers, most famously Bi Sheng, were experimenting with impressing carved and inked blocks of wood onto multiple sheets of paper—the birth of printing.

By then, paper was ubiquitous in daily life in China, from the shopowner to the emperor himself. And thanks to an explosion in trade supported by the government, the world got its first glimpse of paper technology—and coveted it. Japan got its hands on paper technology first, in about 610, from a visiting Buddhist priest.

REAWAKENING

It took the West another 140 years. Supposedly, China lost its monopoly on paper when it also lost the Battle of Talas (in present-day Kyrgyzstan) to a Muslim army in 751. As legend has it, two Chinese prisoners taken in the battle, after no doubt considerable inducement, gave up the recipe.

Within fifty years, huge paper mills were being constructed by the caliphate near the marshes of the Tigris and Euphrates Rivers outside Baghdad, and soon thousands of reams of fresh new paper were being shipped west. By the end of the millennium, thanks to the demand from places like the Royal Mosque of Cordoba, an estimated 60,000 compilations, books of poetry, and scientific texts were being produced per year. Meanwhile, an army of scribes and translators were kept busy converting the older Latin and Greek texts into Arabic. It helped that Andalusian Spain during this period enjoyed a long interval of peace while the rest of Western Europe was hit with a string of natural and man-made disasters, from plagues and poor harvests to the Viking predations.

But this scenario wouldn't last forever. By 1100, Europe had finally emerged into the era of important artistic and scientific achievement known today as the High Middle Ages. It was the confident age of Gothic cathedrals, the beginnings of the scientific revolution, and the Crusades. Even the weather turned warmer.

Meanwhile, the Byzantine Empire, thanks to generations of reduced investment in its armies as well as the cultural stasis of an aged society, was now under assault from every side and would soon break with the West in the Great Schism of the Christian Church.

Even the Muslim Empire seemed to lose its vitality, especially in the West, during this era. Of particular interest to this story was the collapse of Muslim Spain into small competing kingdoms. This at last was the opening the Normans to the north in France needed, and in short order they drove the Muslims out of much of Spain.

Finally Europeans had access to the sum surviving total of their lost cultural and scientific memory. But it had been six hundred years since they'd last had it and so, even with the answers now in hand, even the brightest minds of a resurgent Europe didn't really know what questions to ask. And so, for a century more, the exploration and dissemination of this long-lost knowledge was slow and confused.

Then, in 1203, the seemingly impregnable Constantinople was nearly destroyed from within by an alliance of Crusaders and the Venetian fleet. The Byzantine Empire was indisputably crumbling; and its best and brightest began to abandon it for safer harbors in the West. Soon, Greek and Eastern scholars could be found in the emerging new universities across Western Europe. These scholars knew what treasures were in the library in Cordoba and elsewhere, and they knew in which books their students should look—including the works of the Greek philosophers, Augustine, Isidore, and Cicero's *De Oratore.*

It wasn't long before the race was on across the Continent to be the first to recapture and put to work that regained knowledge. Europe was at last ready to remember.

5
Long-Leggedy Beasties
Memory as Classification

Imagine a dragon.

Chances are, if you live in the Western world (and increasingly the rest of the world as well) you likely picture in your mind a huge, lizardlike creature with a short snout that breathes fire; a long tail with a spearlike end; a pair of giant, leathery bat wings; and possibly a second, smaller pair of wings on the creature's neck.

It is one of the great shared images of mankind. And it pops up almost everywhere, in numerous movies, in television programs, and in literature. It shows up in J. R. R. Tolkien's *The Hobbit*, where the dragon Smaug guards the One Ring and, later, in the *Lord of the Rings* trilogy, where the Ringwraiths ride dragons across the sky in search of Frodo. Most recently, this archetype of dragonhood shows up in the Harry Potter series, the most popular collection of children's books ever written.

One of the most unusual and unforgettable images of this type of dragon appears in yet another classic children's book, Lewis Carroll's *Alice in Wonderland*, this time as the Jabberwock. Carroll's illustrator, Sir John Tenniel, already a famous political cartoonist, had some fun with this fearsome creature by putting it in a waistcoat and giving it the buckteeth and myopic eyes of a Victorian Oxford don (which Carroll, in fact, was). But note, despite these appurtenances, the Jabberwock still retains the conventions of the dragon we all know.

So where did this image of a dragon that we all carry around in our minds come from? What was the source for these very similar images from fantasy novelists to Hollywood spanning more than one hundred years?

In fact, the idea of dragons may be as old as mankind—and centuries old in their current form. The first written stories of dragons come from the Hittites in about 1500 B.C. Dragonlike creatures—Leviathan, the red beast in the book of Revelation—show up in the Bible as well. So, too, do they appear in Greek myth—most famously the Hydra. Add to that the rich history of the wingless but flying snakelike dragons in India, China, and Japan, each with its own unique iconography, and the strange dragonlike worms and serpents that show up everywhere from Scandinavia (the lindworm) to Aboriginal Australia (the rainbow serpent) to the Aztecs (Quetzalcoatl), and you have what seems to be an image of some kind of dragon in the collective unconsciousness of mankind.

This universal appearance of dragons in all of the world's cultures has actually convinced some people that dragons must have once existed in the flesh. But even if dragons have never been real, the question still remains just where the notion of such creatures came from. Theories range from actual dinosaurs surviving into the age of *Homo sapiens,* to exaggerations about existing large reptiles such as crocodiles, large snakes, and monitors, to early attempts to make sense of huge fossil bones.

Probably no single explanation will ever be found. What we do know is that dragons took on very different personas in different cultures. In the Far East, dragons often represent good luck and health. They are also typically benevolent. In Mesoamerica the feathered serpent represents resurrection and knowledge. American Indians saw the dragon as a symbol of wisdom and immortality; to the Australian Aborigines, the dragon ruled the natural world.

But beginning in Sumer and Babylon, and moving across the Levant and into Europe—and then around the world—the dragon was a markedly different creature: dangerous, evil, representing the forces of darkness. Some of mankind's oldest stories involve various Hittite and Sumerian heroes fighting and defeating dragons.

It is a story that never seems to grow old. *Harry Potter* and *Lord of the Rings* aside, consider the run of dragon movies over the last twenty-five years, from *Dragonheart* and *Dragonslayer* to the science-fiction *Reign of*

Fire. Indeed, such iconic films as *Alien* and (ironically) *Godzilla* are essentially variations on the theme of the Western dragon. Dragons also make regular appearances in modern fantasy novels and, of course, in the hugely popular cult role-playing game Dungeons & Dragons.

So, if we are unlikely to ever really know where or when the *idea* of "dragon" arose, is it at least possible to determine where this enduring *image* of the bat-winged, fire-breathing evil dragon came from? Considering it is likely to be the single most important mythical creature in the fantasy menagerie of mankind's collective memory, wouldn't it be nice to know the source of this image?

Well, as it happens, we can almost precisely date that moment: A.D. 1260. And we can see the birth of the modern dragon. (See frontispiece.)

It's a little longer in the tail than the stereotypical dragon we know, though the finlike projections, similar to those found on Chinese and Japanese dragons, are an interesting addition. The extra set of wings, now on the pelvis, is in the wrong place, too. But those are minor differences. What matters is that, unquestionably, this is very much the dragon of our collective memory. It is also astoundingly similar to Tolkien's drawing of Smaug, from its chin whiskers right down to the trefoil tip of its tail.

And yet, this paradigmatic image of a dragon, the wellspring of a million images and a billion nightmares, was created *eight hundred years* before Tolkien put pen to page.

This image of a dragon is, in fact, a hand-painted illustration—an "illumination"—in a thirteenth-century book buried in the medieval collection at the British Library in London. The book itself goes by the distinctly unexciting name of *MS Harley 3244,* and it is one of a class of books known as medieval bestiaries.

As the dragon illustration suggests, these bestiaries are among the most remarkable books ever created. And although few people have ever heard of them, much less held one of these books in their hands, the bestiaries not only are key players in the history of memory, but, as with the dragon, they haunt our memories even today. And as much as the great Gothic cathedrals do, they capture the spirit of the thirteenth century—when the belated arrival of the recorded memories of the ancient world suddenly poured into a Western Europe already regaining its cultural strength. The Middle Ages, begun in darkness and misery, would come roaring out in glory and set the stage for perhaps the most celebrated era in human history: the Renaissance.

A VIGOROUS NEW WORLD

Though they are not generally known, historians often recognize three distinct periods of renewal and prosperity during the seven hundred years of the Middle Ages. The first was the Carolingian era under Charlemagne, that brief light in the Dark Ages. The second, far more obscure, was the even shorter-lived Ottonian dynasty—also known as the "Year 1000 Revival"—in Germany and Italy. It is best known for its production, in "scriptoria" in monasteries at places like Lake Constance, of some of the most beautiful miniatures and illustrated books ever.

But it was the third of these eras, which took place in the twelfth century, that would prove to be the most enduring and influential. The description of this remarkable era by early twentieth-century medievalist Charles H. Haskins still says it best:

> *[It] was in many respects an age of fresh and vigorous life. The epoch of the Crusades, of the rise of towns, and of the earliest bureaucratic states of the West, it saw the culmination of Romanesque art and the beginnings of Gothic; the emergence of the vernacular literatures; the revival of the Latin classics and of Latin poetry and Roman law; the recovery of Greek science, with its Arabic additions, and of much of Greek philosophy; and the origin of the first European universities. The twelfth century left its signature on higher education, on the scholastic philosophy, on European systems of law, on architecture and sculpture, on the liturgical drama, on Latin and vernacular poetry.*[1]

Even Haskins, in studying the era, was astonished by the pace of change—all of it driven by the arrival of waves of ancient texts found in Arabic and Byzantine libraries:

> *The century begins with the flourishing age of the cathedral schools and closes with the earliest universities already well established at Salerno, Bologna, Paris, Montpellier, and Oxford. It starts with only the bare outlines of the seven liberal arts and ends in possession of the*

> *Roman and canon law, the new Aristotle, the new Euclid and Ptolemy, and the Greek and Arabic physicians, thus making possible a new philosophy and a new science. It sees a revival of the Latin classics, of Latin prose, and of Latin verse . . . and the formation of liturgical drama. New activity in historical writing reflects the variety and amplitude of a richer age—biography, memoir, court annals, the vernacular history, and the city chronicle.*
>
> *A library of ca. 1100 would have little beyond the Bible and the Latin Fathers, with their Carolingian commentators, the service books of the church and various lives of the saints, the textbooks of Boethius and some others, bits of local history, and perhaps certain of the Latin classics, too often covered with dust.*
>
> *About 1200, or a few years later, we should expect to find, not only more and better copies of these older works, but also the* Corpus Juris Civilis *and the classics partially rescued from neglect; the canonical collections of Gratian and the recent Popes; the theology of Anselm and Peter Lombard and the other early scholastics; the writings of St. Bernard and other monastic leaders . . . a mass of new history, poetry, and correspondence; the philosophy, mathematics, and astronomy unknown to the earlier mediaeval tradition and recovered from the Greeks and Arabs in the course of the twelfth century. We should now have the great feudal epics of France and the best of the Provencal lyrics, as well as the earliest works in Middle High German.*[2]

Cultural collisions—Alexander conquering the Babylonians and assaulting India, Marco Polo's opening of trade to China, the European discovery of the New World, Commodore Perry's voyage to Japan—are almost always times of great change and innovation. But the twelfth century may have been unique because it was European culture colliding *with its own past.* The result was a four-hundred-year encounter whose implications extend to the present, and reach around the Earth . . . and even into outer space.

In the words of historian of science Lawrence M. Principe, "European scholars embarked on a great 'translation movement' in the 12th century. Dozens of translators, often monastic, trekked to Arab libraries, especially in Spain, and churned out Latin versions of hundreds of books.

Significantly, the texts they chose to translate were almost entirely in the areas of science, mathematics, medicine, and philosophy."[3]

The resulting flood of books that poured into Europe profoundly affected European memory in a number of different ways. One was *completion:* Many of the ancient texts that had managed to survive the centuries in Europe were only fragments of larger texts; the new discoveries completed many of those texts . . . and sometimes transformed their meaning in the process. Another was *expansion*: European scholars of the Middle Ages often saw only one or two works in the corpus of the greatest ancient writers—such as the Greek playwrights—and thus sometimes diminished those figures in relation to other, lesser writers. Finally, there was *substitution*: Europeans naturally learned to admire the best of what they had, only to find what appeared to be superior competitors to those works.

The classic example of this last was the discovery by twelfth-century academics of the works of Aristotle. They had long since learned to exalt Socrates and Plato above all other ancient philosophers—even at the risk of being accused of heresy. The cost of that mind-set had been centuries of single-minded focus on dialogue, governance, and moral decisions.

But the rediscovery of Aristotle, with his concentration on empiricism, classification, and natural science, galvanized Medieval Europe. *Here* was a philosopher whose orientation was perfectly attuned to the new pragmatic and results-oriented Europe of the High Gothic age. In short order, Aristotle became the de facto presiding secular spirit over the era, his grip on the emerging young universities almost complete, and his perspective, while initially liberating, in time becoming its own constraint on innovation.

But that was three hundred years down the road. For the late Middle Ages, the discovery of Aristotle (and Euclid, Archimedes, Thales, and scores of other ancient worthies) was a staggering act not only of recovery of the continent's cultural patrimony but also of unprecedented liberation. Not only were great thoughts and ideas, instructions, and exegeses almost always more impressive than what was already in the libraries of Europe, but as often as not they also contradicted established views. Having been taught to worship the wisdom of the ancients, the more maverick minds of Europe quickly realized that these new texts freed them from the constraints of the status quo.

But first, the thinkers of twelfth century Europe had to get their arms around just what treasures lay in those thousands of "new" volumes. As one might imagine, given the small number of academics in Europe at the time, and the mountain of texts to be explored, this was no easy task. Indeed, it took the rest of the century . . . and even then the work was incomplete and often inaccurate.

THE MEMORIES YOU KEEP

Three immediate tasks faced twelfth-century academics. The first was *selection*. Where should they start? In a culture that was in the process of rebuilding itself, and which was also dominated by the Roman Catholic Church, the choice was inevitable: religion and science.

The second was *translation*. Now that they knew what they were looking for, the researchers had to convert the texts to readable form as quickly and as accurately as possible. This translation triage gave first priority to Latin, as there was no shortage of priests who spoke the language; then Arabic because that was the language of the rewritten texts; and only after that, Greek. This priority system had one very important side effect: It led one of these academics, Leonardo of Pisa, to import Arabic numerals—and thus at last give Europe a system of numbers and mathematics as powerful as its words and writing.

But there was a downside to this translation prioritization as well: Other than Latin-speaking scientists and philosophers, almost everything being translated was already a secondhand translation to Arabic from Latin or Greek—or worse, third-hand from Greek to Latin to Arabic and now back to Latin or the vernacular. Like the childhood game of "telephone," all of this translation and retranslation inevitably allowed a lot of error and confusion to enter into the texts. And in their headlong rush to absorb all of this new information, twelfth-century academics didn't have time to do much fact-checking.

The final joint task was *compilation and interpretation*. Europe was suddenly awash in texts, an ocean of memories—Greek philosophy and mathematics, Roman scientific treatises, ecclesiastical texts recovered from the wilder shores of western Europe, Byzantine histories, Arab medical documents based upon the core texts of the Romans Galen and Ptolemy, regional histories across the continent from the Venerable Bede in the British Isles to the Byzantine court archivists . . . as well as a new kind of long-form poem—the national epic—that was arising from the growing sense of nationalism among Europe's kingdoms. These national epics included France's *The Song of Roland*, the King Arthur stories that begin with Godfrey of Monmouth's *Historia Regum Britanniae*, the definitive written versions of *Beowulf*, Germany's *The Song of the Nibelungs*, and (perhaps not coincidentally) Japan's *The Tale of the Heike*.

Everyone knew the overload situation was going to get worse. Nobles and churches were building private libraries—making the job of scholar or scribe a well-paid profession . . . and thus attracting more talent. A new paper mill had been built in Spain at the beginning of the century, and new ones were popping up all over southern Europe. Meanwhile, the first great commercial empire, the Hanseatic League, was forming across northern Europe. It would ultimately spread goods, knowledge, and accounting books across the continent. And when, a few generations hence, Marco Polo would open the path to China, this trade would go global. Finally, and not least, a wholly new set of social mores—"chivalry"—was sweeping the nobility with its implicit directive to not only be more heroic and virtuous but also be at least marginally literate.

There was so much diverse and unorganized information suddenly available (not to mention the growing demand) that it was almost impossible to simply dive in and explore without getting utterly lost. What was needed was a kind of "metaorganizational" scheme that could be laid over all of this knowledge to organize it by its major themes and then pigeonhole each item within those themes. The two solutions devised by those twelfth-century scholars are still with us today.

The first of these was the *university*. Centers of learning, in the form of cathedral and monastic schools, had been around since the sixth century, acting as a kind of oasis of learning amidst the growing chaos around them and as repositories for books and records. Thus, even from the beginning, the proto-universities were in the memory business.

It would be five hundred more years before real universities emerged in places like Bologna, Paris, Oxford, Cambridge, and Padua. This was not a coincidence: The explosion of new knowledge and memory begun in the twelfth century had suddenly given schools of higher learning a new and much greater purpose. Now their task was not just to educate but to preserve, translate, and investigate.

The students at these universities were no longer taught just the old trivium but a second phase, the *quadrivium*—arithmetic, astronomy, geometry, and music—much of it made possible by the newly discovered texts of the ancients.[4] The best students were even invited to continue on into the study of philosophy and theology—meaning that the very best scholars were not only trained in the higher arts but also skilled in practical science.

The first known university "class" actually took place at the University of Bologna in about 1087 and was taught by a professor named Imerius. His subject was, appropriately enough, that of the Byzantine emperor Justinian's update of Roman law, the *Corpus Juris Civilis,* which had just been rediscovered less than two hundred kilometers away in Pisa, where it had likely sat unnoticed for centuries until the city was conquered by Florence.

But more than teaching these recovered texts, the faculty and scholars at these new "guild" universities (many of them financially supported by royalty in the latest intellectual arms race) were also heavily involved in the recovery process itself. Professors joined monks on the trips to Spain, Rome, and Constantinople in search of lost books, and increasingly it became their task to perform the translations and interpretations of what they found.

In the short term, this made universities the new centers of memory in European culture, enshrining the book, and more particularly the library, as the new heart of university life. This was true both metaphorically and literally. Look at Oxford, arguably the world's first true university: Even today, rising above the city's "dreaming spires" is the dome of the Radcliffe Camera, the circular Palladian masterpiece originally designed not just to hold the university's science library but to stand like a beacon for the library and ultimately the university it represents.

DIGESTING THE PAST

In the twelfth and thirteenth centuries, the primary task of Europe's scholars was to capture and collate all of this newly recovered memory—and there was more than enough to keep every professor, monk, and university student on the continent busy. But what then did you do with all of this digested knowledge? It was too expensive and labor intensive to try to reprint everything in Latin or the vernacular, and there simply weren't enough accomplished readers around to use such books. The next best solution turned out be the other enduring institution: the *encyclopedia.* The twelfth century proved to be the second great Age of Encyclopedias, perhaps even greater than the third, six centures hence. The world was hungry for all of this new knowledge—and impatient to put it to use. And it proved an almost insatiable market for digests, compilations, books of aphorisms, religious histories, lives of saints, natural histories (surveys of mechanics, zoology, and so on) and encyclopedias.

Now, encyclopedias had been around since Pliny's *Naturalis Historia* in the early days of the Roman Empire (he appears to have been editing it when he died investigating the eruption of Vesuvius). And though there were other similar works of writing by Romans in the years afterward, only Pliny's treatise made it through the Middle Ages. Isidore's *Etymologiae* also made it—and would prove even more influential than Pliny's work as a template for encyclopedias. One of the features that made Isidore's work so valuable was that the great scholar, in his race to save the world's memory, had done little transformation on his source material, often just grabbing entire chunks of text from other writers and stuffing them into the *Etymologiae* . . . a boon to scholars five hundred years later because they could read snippets of important, but still-missing works.

There were also newer encyclopedias available, including the giant Byzantine lexicon, the *Suda;* an encyclopedia of science by the Arab physician and pharmacist Abu Bakr al-Razi, and an encyclopedia of medicine by Ibn Sina. As noted, the Chinese had also created *The Four Books of Song,* though it was still unobtainable. And the Ming dynasty, under Emperor Yongle, was about to embark on one of history's greatest feats of memory recording: the 370-million-handwritten-character, 11,000-volume *Yongle Encyclopedia.*[5]

But the recovery of older compilations wasn't the only scholastic work undertaken in this encyclopedic century. It seemed like every university in Europe was also busy preparing whole new encyclopedias and compendia. Many of these new works not only imitated the form of the older works but also borrowed a lot of their content. One of the most commonly used sources, especially for the new natural histories, was Isidore, and it was especially true for the bestiaries that began capturing the popular imagination at the end of the twelfth century.

GOD'S ZOO

The bestiary, or more precisely *Bestiarum vocabulum*, was an extremely popular type of book that enjoyed a brief interval of influence from about 1180 to 1290. Breathtakingly expensive—many would have cost tens of thousands of dollars in today's money—few were ever created. Today, only a couple hundred survive as cherished possessions of some of the world's greatest libraries and museums.

A simple description of a bestiary is that it is a book that combines pictures of animals from around the world with descriptions of their behavior and their role in the larger ecosystem.

But that doesn't begin to describe the experience of actually reading one of these volumes. The best of them contain dozens of exquisite miniature paintings, many of them masterpieces. As for the animals themselves, this isn't some modern field guide: Rather, the selection of creatures ranges from the prosaic (mouse, cat, and dog) to the fabulous—unicorn, phoenix, manticore, and, of course, the dragon. Indeed, the very fact these mythical creatures continue to survive in our common memory and are so deeply embedded in our culture that they are likely to survive for at least another millennium is in large part thanks to these bestiaries.

But there is much more. For modern man, reading a medieval bestiary can be a profoundly disorienting, even disturbing, experience.

This is the medieval mind at full flowering, a glimpse of how sophisticated and elegant a society can be constructed without the rigors of experimental science. The epistemology that underlies the bestiaries is as complex as any modern scientific taxonomy, and the metaphysics of the world it portrays are as subtle, irrational, and counterintuitive as anything found in string theory or particle physics.

Just as the modern mind often conflates the early Middle Ages (the Dark Ages) with the later, so, too, it often mistakenly assumes that the people of the Middle Ages were simple and ignorant. That is, as stupid serfs at the mercy of a brutal nobility. But medieval society was as multilayered (if not as specialized) as our own. And to dismiss that world as solely one of fat, stupid, and drunk farmers dancing and groping in a Breughel painting; otherworldly and self-obsessed clergy; and haughty, overjeweled nobles in tights and codpieces is to ignore their present-day counterparts at NASCAR races, Hollywood celebrity events, Congressional subcommittee hearings, and university faculty meetings. Knowledge changes, but intelligence does not. Medieval man, though usually less literate and numerate than modern man, had a mind, and an imagination, as powerful as our own.

Indeed, over the last two centuries, a number of important writers, artists, and philosophers—Henry Adams, G. K. Chesterton, the Pre-Raphaelite painters, and Henri Bergson to name a few—have suggested that we have lost a lot with the rise of the modern industrial world and that it is quite possible that for all of the difficulties of that era, medieval man had a richer inner life, a deeper and more comfortable relationship with the natural world, and most likely was even *happier* than we are.

But entering into the medieval mind, experiencing the world through those eyes, and dreaming the dreams of a man or woman of the Middle Ages grow ever more difficult by the year. The pace of life has accelerated almost continuously for the last eight centuries, and now—thanks to the World Wide Web and the microprocessor—has begun to move at a pace beyond any biological system, including the human brain. In our world, a single image may draw fifty million viewers in the course of an hour on the Web, our telescopes can look back 6 billion years to the beginning of the universe, and we plot the trajectories of particles so small that they compare in size to a human being as we do to the solar system.

We also live in a universe where God, if He exists at all, has retreated to the dimmest corners of the Heisenberg uncertainty principle and the cosmological constant. Thanks to quantum physics, we are no longer sure about the existence of anything, much less able to describe it with certainty. Once-obvious truths, such as causality, are now gone. In fact, the very notion of truth itself is under assault from every side. Darwinian selection and genetics tell us that human beings are merely a chemical accident. We are a tiny mistake in a far corner of a universe that we probably won't ever understand, and that itself seems to have no meaning and no purpose.

But for medieval man, humanity, at the indulgence of a watchful

God, stood at the very center of the cosmos. Man was the master—and custodian—of all living things. Even the sun and stars wheeled around him, their orbits as perfect as the mind of the Creator Himself. Moreover, all of Nature was a vast puzzle palace created for man to decipher God's will—a theater in which every creature exists to offer lessons, good and bad, about God's grace, Christ's salvation, and of the Resurrection.

This was the world of the bestiaries. Because of their rarity, their beauty, and most of all their strangeness, they have long been the subject of academic research and speculation. Yet there are still huge holes in our knowledge about them.

For example, we don't really know who invented them, who owned them, even exactly what they were used for. It had long been assumed that they were created by artisans for very wealthy families (or groups of neighboring families), mostly in England but also in France. In this theory, bestiaries were a combination of prestige object, travel book, cabinet of curiosities, and theological instruction book. They might be displayed to impress visitors, entertain adults, and instruct children—via stories and images that both thrilled and delighted—on the importance of becoming good Christians.

But there is a second, newer theory that holds that few families—even the wealthiest nobles—would have been willing to pay the price for such a volume, and that the bestiaries were instead commissioned by the wealthiest abbeys and monasteries. There they would have been used, perhaps, for the moral instruction of aristocratic novices and the entertainment and edification of the priests and monks.

The truth probably lies somewhere in the middle. The bestiaries were hugely expensive, and thus their prospective audience, in a world without much of a commercial class, would have been restricted to a very small number of very powerful people, be they noble families or large institutions—the way, say, a private jet is today.

NATURE'S PRIMEVAL REALM

Novelist T. H. White, best known for his sequence of novels about King Arthur, *The Once and Future King* (the basis for the movie *The Sword in the Stone* and the musical *Camelot*), was also an important medievalist. And he was entranced by medieval bestiaries—especially the copies at his alma mater, Cambridge. In 1954 he published the only book on the subject ever written for the average reader, *The Book of Beasts*, about the best of the Cambridge bestiaries, "MS. li.4.26."[6]

Besides being one of the finest-illustrated—and, thanks to White, best known—of all of the bestiaries, the MS. li.4.26 (hereafter the *Cambridge Bestiary*) is also one of the few whose origins can be fairly accurately placed. Its story is a glimpse of what the memory business was like at the end of the twelfth century.

As tradition tells it, the *Cambridge Bestiary* was created at the Abbey of Revesby in the North Midlands—a region in north-central England that today has no official status but is usually considered to include the northern parts of Derbyshire and Nottinghamshire (the latter of which has its own historic resonances: Robin Hood reputedly began operating out of Sherwood Forest not long after this bestiary was written).

White believed the *Cambridge Bestiary* to have been created about A.D. 1130. He based this date on several features of the book, including the appearance (and nonappearance) of some animals in its pages, and the apparent transition in its pages from illuminated to nonilluminated initial letters—the former technique forbidden at that time by the abbey's very austere order, the Cistercians.

That would put the book's creation during the reign of the last Norman king of England, Stephen. But later scholarship has moved that date later, to the end of the twelfth century. That would put it during the reign of a much more famous king, Henry II—one of the greatest English monarchs, the first Plantagenet, the husband of the legendary Eleanor of Aquitaine, and the man who ordered the death of Thomas à Becket. Some contemporary researchers suggest that the *Cambridge Bestiary* might even have been written as late as the reign of Henry's son, the less impressive but even more famous Richard the Lionheart.

Those illustrious names may provoke specific images in the mind of the reader: crenellated castle walls, elegant ladies in waiting, colorful and rowdy jousts, and, perhaps, Errol Flynn in green tights. The reality was much different. White quotes the Edwardian historian George Macaulay Trevelyan:

> *What a place it must have been, that virgin woodland of old England, ever encroached on by innumerable peasant clearings, but still harboring*

> *God's plenty of all manner of beautiful birds and beasts, and still rioting in a vast wealth of trees and flowers. . . .*
>
> *In certain respects the conditions of pioneer life in the Shires of Saxon England and the Danelaw were not unlike those of North America and Australia in the nineteenth century—the lumberman with his axe, the log shanty in the clearing, the draught oxen, the horses to ride to the nearest farm five miles away across the wilderness, the weapon ever laid close to hand beside the axe and the plough, the rough word and ready blow, and the good comradeship of the frontiersman.*
>
> *. . . Every one of the sleepy, leisurely garden-like villages of rural England (today) was once a pioneer settlement, an outpost of man planted and battled for in the midst of nature's primaeval realm.*[7]

This may sound idyllic, but there's much more. This was not too much longer than one human lifetime after William the Conqueror, the Norman invasion, and the slaughter and purges that followed. England still felt like an occupied country, the Anglo-Saxon natives seething under the rule of a French-speaking nobility.

Thanks to the disputed succession—and the subsequent weak leadership—of Stephen, England collapsed into a civil war called, tellingly, "The Anarchy."

The Anarchy is the world of the Brother Cadfael novels: nobles switching loyalties back and forth, forcing their serfs into service as soldiers and for sieges, wholesale slaughter, and torture. According to Trevelyan, entire districts in the country were "depopulated." A few miles away from the Abbey of Revesby, in the Fenland, Geoffrey de Mandeville and his army were living off the countryside, plundering, raping, and murdering everyone in their path.

In Peterborough, a monk whom White suspects may have known the authors of the *Cambridge Bestiary* captured the horrors of the era in the last, tragic entries to the *Anglo-Saxon Chronicle*, the great history first commissioned by King Alfred four centuries earlier:

> *They greatly oppressed the wretched people by making them work [building] castles, and when the castles were finished they filled them with devils and evil men. Then they took those they suspected to have any goods, by night and by day, seizing both men and women, and they*

> *put them in prison for the gold and silver, and tortured them with pains unspeakable, for never were any martyrs tormented as these were.*[8]

This is a world of random violence, social chaos, and desperate prayers for salvation: to the local lord, to the king, and most of all, to God.

Thus, in picturing the Abbey of Revesby, we should be imagining less a quiet and ordered place—its peaceful monks toiling away, day after day, to the precise schedule set by the prayer bells—and more a walled bastion, the men struggling through long hours to create products that would earn the abbey money and save their souls in the process. The monks creating the *Cambridge Bestiary* were probably less contemplative than vigilant, even as they worked regularly scanning the horizon for men in helmets or on horseback.

A CIRCLE OF SCRIBES

The actual production of the *Cambridge Bestiary* and other bestiaries is also different from what most people might imagine it to have been. The lone monk scribe toiling away in a tiny cell is mostly a myth. Producing bestiaries was a business, for which an abbey could take advantage of its intellectual capital—literate men with artistic skills and a lot of time on their hands—and turn it into operating capital. Given that the abbey could earn the modern equivalent of $50,000 to $200,000 per book, all with a labor overhead of essentially zero, it's not surprising that production was less like a solitary craftsman and more like a proto–assembly line.

White suggests that most bestiaries were created by dictation: "Since books could only be duplicated by hand, it was reasonable to duplicate as many as possible at the same time, by dictation. Several scribes seated at their desks round the Scriptorium could produce a limited edition as the text was read out to them."[9]

Among the evidence in support of this argument is that the errors that regularly crop up in the bestiaries are typically those—"*eximiis*" instead of "*et simiis*," is White's example—that most probably result not from a misreading, but a *mishearing.* There is also the fact that the surviving bestiaries can be neatly lumped into a handful of "families"—that is, groups of three or four that feature similar narratives and images.

The notion of a single author is also wrong for many of these books.

White notes that the *Cambridge Bestiary* appears to have been written by *two* scribes, whose handwriting differed slightly. One can almost picture dictation teams working in shifts, those not writing perhaps working the fields or cooking meals.

This image of an early assembly line is further supported by the fact that the writers of the bestiaries had no part in the illustrations. That work was done by the "limner"—the artist. We don't know for certain which came first, the pictures or the words—whether the scribes wrote the text leaving holes to be filled later with pictures, or vice versa. But certainly there must have been rules about the apportionment of real estate on the pages. That there are often divergences between the descriptions of animals in the text and the way they appear in the accompanying illustration suggests that communication between the two parties was probably rare, and competition was not unknown. And in keeping with the notion of mass illustration, it is also important to keep in mind that while many of the illustrations are superb, occasionally approaching genius, they are not artistic masterpieces of the highest order like those found within the next great publishing fad, the Book of Hours—a notable example being that of Jean, Duc du Berry, a masterpiece that took a century, and cost a true royal fortune, to create.

No, the greatness of bestiaries lies in the fact that they aren't the creation of specialized court painters and isolated academics but skilled craftsmen immersed (whether they liked it or not) in the daily life of the society around them. Those same monks and illustrators likely milked the cows that morning before settling down at their easels, and might have spent part of the afternoon out in the fields. Moreover, as wealthy as their clients might be—whether it was an abbey or a wealthy noble family—the authors and artists of the bestiaries knew they were creating for a popular, rather than a scholarly, audience. This led them, to the benefit of readers then and now, to give barely hidden priority to the entertaining over the edifying, and the sensationalistic over the educational.

It was not an easy life, but by the standards of twelfth-century England (and France), it was a bearable one. And though their work would remain anonymous, the monks must have taken satisfaction that they were producing work of the highest order in the eyes of both man and God.

Imagining this world sends White into a reverie of impressions:

> *Seated at their high desks like the professional craftsmen that they were; with quill pens and knives to trim them; with iron-gall inks, on skins which were difficult to prepare; with the strong downward strokes of their 'I's . . . with their attention concentrated upon the lector who dictated to them, or not concentrated, according to the mood of the moment; with no pauses to indicate the pauses of the dictation; with their words running over from one line to the next or*
> *b e i n g s t r e t c h e d o u t*
> like this to fill up a space; with their quotations from scripture, which, if well known, were merely indicated by the initial letters of the words; immured in not comfortable barracks, since the Cistercians were a self-denying order; surrounded by the dangers of a civil war and the difficulties of frontier life; with their tongues between their teeth and their blunt, patient, holy fingers carefully forming the magic of letters . . . [10]

THE BEST BESTIARY

Arguably the greatest of the bestiaries is called *Bodley 764*. Probably created in Salisbury, it appears to be a cousin of the *Cambridge Bestiary* that entranced T. H. White.

It isn't the most beautiful of the bestiaries—in fact, it isn't even the most beautiful at Oxford: Faculty members sometimes jokingly referred to it as "our second-best bestiary" after the older and flashier one at the Ashmolean Museum nearby.[11] But in terms of completeness and the overall quality of illustrations, its balance between text and art, and most important, its immeasurable influence on literary history, *Bodley 764* has no equal among the bestiaries. It is the sourcebook for much of our modern fantasy.

There is a wonderful, almost supernatural symmetry to *Bodley 764*. For one thing, as an ultimate creation of the encyclopedic aspect of its era, it has made its home for the last six hundred years within the ultimate manifestation of the university aspect of that same era: the Bodleian Library at Oxford.

And if, unlike the dragon in *MS Harley 3244*, it features no single illustration that we can point to and say *that* is the source of some modern iconic image, *Bodley 764* is even more important in its influence on human memory. Residing as it did in the "Bodley," it served as the portal into the world of bestiaries for generations of students and academics—among the latter being Professors Charles Dodgson (Lewis Carroll), J. R. R. Tolkien, and C. S. Lewis.

If you wanted to list the prime sources of modern fantasy, you wouldn't have to look much further than *Alice in Wonderland* and *Through the Looking Glass, The Hobbit* and the *Fellowship of the Rings, The Chronicles of Narnia*—and, to include Lewis's most famous descendant—the Harry Potter books. In other words, the multiplicity of modern, nearly identical-looking dragons, phoenix birds, and other mythical creatures in the modern world is no coincidence. Nor is the fact that, again with perfect symmetry, *Bodley 764* can be found in the oldest part of the Bodleian: Duke Humfrey's Library. The first "modern" library, it was donated by the brother of King Henry V in 1487 and was used in the filming of the Harry Potter movies.

Held in one's hand, *Bodley 764* is a normal-sized, unimpressive, leather-bound book, and in heft surprisingly light given that it is both made of animal-skin parchment and acknowledged as the most complete bestiary in the British Isles. Despite being most complete, it is still missing its title page, which has denied generations of researchers the name of its original owner.

Martin Kauffmann, the librarian who watches over *Bodley 764*, describes it as being "almost alive."[12] Thanks to its parchment pages, the book swells slightly in size and its pages soften on humid days; then it becomes stiff and bristly when it is dry.

Leafing through the book is to fully appreciate that *Bodley 764* is indeed memory made flesh. The parchment pages are shockingly supple for their age, and one can feel the crisp face of one side of each sheet and the soft, slightly fuzzy side of the other. In a world of photolithography it is sometimes easy to forget that each of the exquisite miniatures within is in fact an original painting, the work of an artist's hand at the time of Crusades. Only the luminescent colors, made from ground jewels, and the gold-leaf highlights, brighter than any reproduction, remind the viewer that these pages were written nearly two hundred years before Gutenberg.

The illustrations themselves, and the narratives that describe them are an entertaining, eccentric, and sometimes confusing mélange of accurate protoscientific observation, sheer fantasy posited as reality, and an overarching (and often, to modern eyes, maddening) need to present everything in the natural world as a lesson devised by God to edify mankind. And while it can be interesting to find mermaids described in the book as naturally as geese, or to see boar tusks sprouting out of the trunk of an elephant, what is most surprising is to find the behavior of common European creatures so inaccurately described. Thus the pelican draws blood from its own breast to bring its dead young back to life; bears lick lumps of flesh into cubs; hedgehogs catch falling fruit on their spines; and the salamander crawls through a fire unburned. The modern reader can't help but ask: How did they get these everyday events so wrong? Didn't they *see*?

The fact is, they saw what they wanted to see—as do we. And the world they wanted to see was one suffused by the glory of a loving but stern god. The descriptions they chose to believe in the face of real-life evidence were those of the ancients: much of them derived from Isidore's *Etymologiae*. Eight hundred years before, Isidore had set out in a desperate attempt to save the world's memory, and now his dream—mistakes and all—had been realized.

Medieval men and women believed in absolute truth, in the Word of God, and in redemption. Those beliefs gave their lives meaning, they provided succor in the face of the world's hardships, and especially offered hope of a better world beyond. In that world, even the lowliest person mattered: After all, hadn't God filled the Earth, sea, and sky with creatures to bring the Word to *all* men, just as Christ had died for all of mankind's salvation? You had only to decipher the messages delivered in the world around you, as explained by a book like the bestiary, and you could find the answer, you could learn God's plan for you. To the medieval reader, the bestiaries were a pathway to mankind's memory of God. For modern man, the bestiaries are a unique glimpse into the memory of medieval man.

If a single book seems a fragile vessel for us to sail upon to another world, keep in mind that *Bodley 764*, like its counterparts, has already managed to survive nearly a millennium. Entire empires have risen and fallen in that time; buildings constructed to last forever have crumbled to ruin; even some of the stars the scribes looked at in the sky as they huddled off to vespers have dimmed in that time. Yet *Bodley 764* still looks fresh and bright with color—probably better than the book you bought last week at the mall. It may well survive as long as our dreams of dragons.

THINKERS' THINKERS

The bestiaries weren't the only new phenomenon created at the nexus of encyclopedias, universities, and faith. The early thirteenth century also saw the rise of a new form of learning: *scholasticism*.

Scholasticism was a dialectical system; it, too, was derived from the ancients—in this case, the Athenian Academy. Like Plato's school, it thrived on argumentation, rational thinking, and precision; and like Aristotle, it sought logical and self-contained systems that could encompass all knowledge. This new philosophy swept through European universities so completely that its echoes are still heard today in college lecture halls and the "Socratic method" of teaching.

Scholasticism had been around, in nascent form, since the eleventh century, when leading scholars such as Peter Abelard and Archbishop Anselm of Canterbury had been inspired by the first translations of Greek philosophy coming out of Ireland (where the monks had been the last to preserve the language in the British Isles).

But scholasticism reached its full flowering in the thirteenth century when a flood of translated Greek works—most notably Euclid's *Elements*, with its rigid logic—reached the universities of Europe. There they were read by the scholastics, among them some of the most powerful thinkers in history: Duns Scotus, William of Ockham (from whom we get "Occam's razor"), Saint Bonaventure, and greatest of all, the Italian Dominican priest Thomas Aquinas. And it wasn't only among Christian scholastics that these translations proved a revelation: In Spain, they were read by a young rabbi and physician named Moses ben-Maimon: Maimonides.

For each of these scholars, the Greek philosophers offered a vision of seeing and organizing the world in a new way: logical, rational, and precise. Free of all of the ghosties and ghoulies and long-leggedy beasties of the medieval world. Self-contained and intellectually impregnable. And all for the glory of God. Scholasticism was, in the end, not an act of radical revolution but reactionary purification. The scholastics set out to prove that not only did God exist but that His universe was both rational and comprehensible to the mind of man.

It was one of the most difficult intellectual challenges ever undertaken—and the results remain both staggering in their ambition and stunning in their achievement. Maimonides's fourteen-volume *Mishneh Torah,* his codification of Jewish law, still carries canonical authority in the Jewish faith. That's why he's called the "haNesher haGadol" (the Great Eagle) of Judaism.

Aquinas's influence may have been even greater. His *Summa Theologica* was recognized even by his contemporaries as the zenith of scholasticism, and it remains the single greatest intellectual edifice of the Roman Catholic Church—indeed, the doctrines of Saint Thomas are the Church's own. In this amazing work, Aquinas seemed to gather together all of human memory to that date, insert it neatly into a great logical matrix, and tie it to the will of God. And like the brilliant intellectual craftsman he was, Aquinas dovetailed it all together so seamlessly that it was hard to find even the tiniest opening to attack it. One can even credibly make the claim that all Western philosophy (and certainly theology) since the publication of the *Summa Theologica* has been one long argument against that work.

THE END OF CERTAINTY

It can be a sad experience to speculate on what might have happened if this world of universities and encyclopedias had continued to flourish, and if the medieval world had evolved into our own along a more linear path. Would we have still had a scientific revolution but also somehow retained the richness and majesty of the Middle Ages? Would we have found a better way to integrate faith and reason? Would we still have all of the technologies that improve our daily lives while also enjoying the honor, the romance, and the courtliness? Would we still be looking at the natural world filled with both wonder and purpose—and the comfort of an awaiting heaven—that has been lost?

We will never know, because it all ended within a matter of decades. The fourteenth century would prove to be one of the most horrible and destructive periods humanity has ever known: The Byzantine Empire collapsed, as did the Mongol Yuan dynasty in China. The Catholic Church split into three parts, each with its own pope. England and France entered into the devastating 100 Years War and, thanks to the beginning of the Little Ice Age, famine killed millions.

And all of that was merely a sideshow. The worst arrived in 1347: the Black Death, killing one-third of the population of Europe.

The Europe that emerged after this series of catastrophes was very different from the one that entered it. Unlike in the fall of Rome, this time mankind's memory had not been lost; it endured not only in the minds of survivors but also in the sturdy books that now filled royal, ecclesiastical, and private libraries throughout Europe and Asia. This time, mankind would come back—and fast.

But when it did, things had changed. God had proven Himself to be more inscrutable, and perhaps less amenable to rational thought, than the scholastics had imagined. This time, when Europe came back, its philosophy would be centered on man himself. The new goal would not be to suffuse the universe with faith, but with science.

6
Theaters of Memory
Memory as Reference

On February 17, 2000, hundreds of people gathered in the Campo de' Fiori ("the field of flowers")—a large square, or piazza, in Rome beneath the statue of a man dressed in monk's robes, the hood casting his face in eternal shadow.

The crowd, a motley collection of anarchists, atheists, freethinkers, and pantheists, quickly set up booths and speaking platforms, plugged in loudspeakers, and covered tables with pamphlets . . . and proceeded to denounce the Roman Catholic Church to whomever walked past.[1]

The crowd, different in their beliefs, but united in their common hatred, had gathered on this particular day to celebrate an event that had occurred four hundred years earlier: the public burning of a heretic. And the man of the hour on both occasions was the brooding figure on the nearby pedestal: the Dominican friar, theologian, philosopher, and scientist Giordano Bruno.

Passersby could be forgiven if they didn't recognize the object of veneration. By the time he faced his auto-da-fé, Bruno was already almost a forgotten figure, having spent the previous seven years imprisoned in the center of the city in the Tower of Nona while he underwent a seemingly endless trial, conducted by Galileo's inquisitor, Cardinal Bellarmine. In the end, he was found guilty of a long string of crimes against the Church including holding erroneous opinions about Christ, the Incarnation, the Trinity, Transubstantiation, and the Mass; a disbelief in the virginity of Mary; and, most uniquely, a belief "in a plurality of worlds and their eternity."[2]

The same passersby might also be forgiven if they wondered what the fuss was all about in the year 2000: Not only were Bruno's crimes not too far from the beliefs held by many modern Catholic theologians (and almost a prerequisite for becoming, say, the Archbishop of Canterbury), but just days before the demonstration the Vatican itself had publicly apologized for Bruno's burning.

AN INDOMITABLE WOMAN

For most of those four hundred years that passed since Bruno's ashes were dumped into the Tiber River, he was all but forgotten—one of those minor figures of the Renaissance who paid the ultimate price for being ahead of his time (and not too far ahead either: His was one of the last burnings of a heretic by the Church). And Giordano Bruno might have remained obscure had not a reader in Renaissance history at the University of London, the middle-class daughter of a shipyard engineer from Southsea, Portsmouth, decided to look deeper into Bruno's newly rediscovered works on memory.

This was 1951. Just eleven years earlier, Bruno's name had briefly resurfaced when the Vatican announced, in an academic paper, that it had discovered the long-lost transcripts of Bruno's trial. Three years later, in 1943, the author of the article, a Cardinal Giovanni Mercati, publicly announced that based on those transcripts he believed that Bruno was indeed guilty of his accused crimes. But there was a war on and, moreover, Italy was the enemy, so interest in the story was slight and quickly forgotten.

Except by Frances Yates. In retrospect, it seems natural that this no-nonsense British scholar (she professed to never seeing much practical value in the mystical arts practiced by her subjects) might find a common ground across the centuries with this radical monk of the Italian Renaissance. Both were mavericks, staking out intellectual positions at odds with the common view. Both were also internationalists: Bruno spent years wandering about Europe, accepting shelter and sanctuary from any prince or university that would take him. Yates, who had seen her brother die in World War I and herself survived the Blitz in World War II, had grown bitter about the costs of nationalism and division—and began to devote her life and career to interdisciplinary studies and transnational politics.

Most of all, they were fearless. Bruno's life was one long battle against the established order and its unchallenged beliefs, his horrible death the inevitable consequence. He might have survived with an early and carefully worded recantation. But instead, he forced the Inquisition to demand a full recantation . . . and refused to give it. When, at last, as his sentence was named, Bruno reportedly made a threatening gesture at his judges and then spit at them the words, "Perhaps you pronounce this sentence against me with greater fear than I receive it."[3]

The situation wouldn't be life or death for Frances Yates as she began to dig deeper and deeper into the increasingly strange nature of Bruno's writing. But what she encountered there—and its implications for the official history of the Renaissance—was so unsettling that to publish it risked giving her a reputation as a crank . . . and for a woman scholar in the early 1950s, that would have been a career killer.

But Yates didn't hesitate. And in 1964, she published *Giordano Bruno and the Hermetic Tradition*. She followed up two years later with the work for which she is best known, *The Art of Memory*.

The first book stunned the world of medieval historians. The second sent shock waves through the academic world and even lapped up against the mainstream reading audience. And what shocked these audiences was exactly what had unsettled Yates in the first place: the notion of a second, hidden history of human thought that was secretive, antirational, built upon magical thinking and a quest for godlike omniscience . . . and all constructed around the art of superpowered memory that had briefly surfaced with Cicero and the orators of the late Roman Republic.

It was an idea so fantastic that Yates herself had initially fought it. But the deeper she searched, the more the conclusion became inescapable, not least because it explained so many mysteries of medieval intellectual history. Wrote Yates:

> *I could not understand what happened to the art of memory in the Middle Ages. Why did Albertus Magnus and Thomas Aquinas regard the use in memory of the places and images of "Tullius" [Cicero] a moral and religious duty? . . . Why, when the invention of printing seemed to have made the Great Gothic artificial memories of the Middle*

> *Ages no longer necessary, was there this recrudescence of the interest in the art of memory in the strange forms in which we find it in the Renaissance systems of [Giulio] Camillo, Bruno and [Robert] Fludd?*[4]

For Yates, the biggest question of all was: How did a practical tool to help orators improve their rhetorical abilities become transformed—out of sight of the world—into a metaphysical and ethical system that got its practitioners burned at the stake for heresy?

The search for answers took Yates into the most unlikely historical back alleys. And, incredibly, the strangest of all turned out to be the most important: *Hermeticism*.

THE SECRET KNOWLEDGE

Hermeticism was an occult religion that first appeared in about the second century A.D.—but claimed to date back to ancient Egypt—as a sort of last-ditch attempt to salvage traditional pagan religion in the face of the growing influence of Christianity. There was a lot of this going around at the time; it's no coincidence that this period also saw the rise of the equally heretical (and persistent) Gnosticism.

In its original incarnation, Hermeticism accepted some of the key tenets of the Judeo-Christian tradition, notably the goodness and perfection of a single God and the need to purify oneself of sin . . . but after that, everything went very strange indeed. In particular, as the name suggests, Hermeticism used pagan rituals (notably image worship) to venerate a figure named Hermes Trimestigus—"trice-great Hermes"—who appears to have been a hybrid (a "syncretic") of the Greek messenger god Hermes and the Egyptian god of intelligence Thoth (he's the one with the head of an ibis or baboon). According to the tenets of Hermeticism, Hermes Trimestigus was the pathway to the great corpus of "secret" knowledge that included everything from alchemy to the secrets of the universe.

Unsurprisingly, one of the other characteristics of Hermeticism was its mistrust of rationality—not a bad position to hold when you are convinced you can know the mind of God. What *is* surprising to most modern readers is that almost all of us encounter the primary symbol of Hermeticism regularly in our daily lives. It is the *caduceus*, the two snakes winding around a winged pole that is the most common symbol for commercial health-care organizations in the Western world (and is itself the result of confusion with another snake-around-a-stick symbol, the Rod of Asclepius—it's a complicated story). This same caduceus, with or without snakes, is traditionally seen in sculptures in Hermes's hand.

Hermeticism is one of those classic pseudoscientific belief systems—the most famous being astrology—that arose before modern empirical science, and that have shown an amazing ability to survive in the face of it, often by going dormant and reappearing centuries later. In the case of Hermeticism, during its first century it managed to pop up in Syrian, Coptic, Arabic, Armenian, and Byzantine Greek versions. Even Augustine took time to attack it. And then, after disappearing in Europe for the late Middle Ages, it reappeared during the Renaissance. Frances Yates was able to trace this reemergence pretty precisely to 1460 and an agent of a Tuscan ruler, the legendary patriarch of the de Medici dynasty, Cosimo.

Cosimo de Medici, like many of the rulers of the era, was on the hunt for ancient manuscripts, and he had sent out his agent, known only as "Leonardo" to search the old monasteries of Europe to find them. One of the prizes Leonardo returned with was the *Corpus Hermeticum*, the core text of Hermeticism—a reminder that medieval and Renaissance scholars, in preserving the past, saved both the good and the bad . . . and mistakenly venerated both.

From there, it was translated by a member of the de Medici court named Marsilio Ficino, who published it in 1471 as thirteen short treatises ("tractates"). And from there it fell into the hands of Giordano Bruno.

The great figures of the Renaissance always seem positioned at the confluence of multiple streams of memories flowing from a newly rediscovered past and as recipients of the latest discoveries coming from the emerging world of science. Think of Johannes Kepler and Galileo inheriting both the rediscovered works of Ptolemy and Aristotle's *On the Heavens*, as well the theories of Copernicus and the research of Tyco Brahe. Or William Harvey, working from (and eventually disputing) Galen while drawing on the work of anatomists of the previous four centuries—from Ibn al-Nafis in the thirteenth century to Michael Servetus a few decades before—to discover the circulation of blood.

Giordano Bruno was no exception. Like a certain type of brilliant man or woman throughout history, Bruno was attracted to the notion of secret knowledge and the awesome power reputedly wielded by its practitioners. And Hermeticism was believed to be the deepest and oldest form of magic, drawn from the darkest corridors of memory dating back to the pyramids and the Sphinx. But Bruno was also a man of the cloth, a monk, charged to defend the Catholic faith, the church, and its tenets. And needless to say, the Holy See—and its enforcement arm, the Inquisition—did not look kindly upon practices such as alchemy and magic, and was willing to use the tools of torture and death to root it out of the population.

No fool, Bruno well understood that to even read the *Corpus Hermeticum* was heresy, and a possible death sentence. But being a maverick thinker, he couldn't help but be drawn ever deeper into Hermeticism and other forbidden areas of knowledge, such as the anti-Trinitarian Arian Heresy and the humanistic Protestantism of Erasmus. And try as he might to hide these interests, in the close confines of the monastery they occasionally slipped out . . . and were noted. At different times he was caught throwing away images of saints or suggesting questionable readings to novices. Worst of all was the discovery—hidden in a privy—of his annotated copy of a book by Erasmus.[5]

Facing a hearing by the Church, Bruno fled Naples, and for the next seven years he was on the move—first throughout Italy, eventually resting in Venice, where he published a book, then on to France and ultimately the sanctuary city of Calvinist Geneva. There he abandoned his monk's robes and took to wearing everyday clothes so as not to be recognized.

Throughout these journeys, Bruno did not dare carry any books or writings that might betray him to the local authorities. Instead, he carried a secret weapon in his head: the Art of Memory. This was the second ancient stream that had reached Giordano Bruno at a young age and changed his life forever. Even as a youth, Bruno had been noted for his powerful memory. But it was *Ad Herennium*, Quintilian, and Cicero's *De Oratore* that fired Bruno's ambition to become a master of the Art of Memory as great as the Roman orators.

But whereas the ancients had almost exclusively seen "synthetic" memory as a tool of their trade, Bruno (as was his way) saw something else . . . something mystical. After all, wasn't the Art of Memory a means for mere mortals to access superhuman quantities of knowledge? And as they did so, might they draw parallels, connections, and distinctions far beyond those made in everyday life? And wasn't this like the mind of God?

Even more than that, it wasn't lost on Giordano Bruno that many of the disciplines of Hermeticism—notably astrology—were themselves kinds of memory "theaters" in which complex meanings were attached in a mnemonic way to a fixed system of symbols. It was Yates herself who noticed that, in a famous description from *Ad Herennium,* there are not just any testicles hanging from a statue's cold stone hand, but a *ram's* testicles—in other words, Aries, the very first sign of the Zodiac—suggesting a much older connection between the memory arts and astrology than previously thought.[6]

Such a connection between astrology and the Art of Memory seems natural, and no doubt Bruno saw it the same way. Now he could embed the secret knowledge of the universe into the familiar wheel of the Zodiac and store it all in his powerful brain where no one could find it. Or so he thought. So powerful was this new art that Bruno credited it with its own mystic powers. And since Hermeticism was a syncretic belief system—that is, one that readily absorbed new beliefs—it soon added the Art of Memory to its tradition.

It wasn't long before the notion of memory theaters captured the imagination of progressive thinkers and mystics across Europe. Some of the more ambitious even decided to convert this mental technique into real physical structures. That is, actual private walk-in theaters of memory filled with all sorts of unusual structures and items on which to hang pieces of synthetic memory. There were probably two reasons for doing this: First, if the design was already memorable and painted with "blood," it would make the memorization process easier by going halfway; and the process of memorizing forbidden, and potentially fatal, knowledge was not something you wanted to do in public.

MEMORY AS MAGIC

Probably the most famous (if only because Frances Yates devoted many pages to it) of these Renaissance physical memory theaters was built by the aforementioned Giulio Camillo in Padua in about 1530. It instantly made him one of the most celebrated figures on the continent. In a letter to Erasmus, a certain Viglius Zuichemas, after noting that everyone was talking about Camillo and his room crowded with images, said:

> *They say that this man has constructed a certain Amphitheatre, a work of wonderful skill, into which whoever is admitted as a spectator will be able to discourse on any subject no less fluently than Cicero. I thought at first this was just a fable until I learned of the thing more fully from Baptista Egnatio. It is said that this Architect has drawn in certain places whatever about anything is found in Cicero . . . Certain orders or grades of figures are disposed . . . with stupendous labour and divine skill.*[7]

The next time Zuichemas writes to Erasmus, it is after Camillo has given him a tour of the memory theater—and he can hardly contain his excitement at the experience. It is from this letter that history gets its first description of the structure.

The theater proves to be a small structure—perhaps the size of a small modern suburban bedroom, large enough to hold two visitors at a time. Organizationally, it is laid out in a series of rings, segmented into pie-shaped sections. Each of these sections corresponds to the known planets, and each ring corresponds to a common symbolic image from mythology, such as the sandals of Mercury or the Gorgon sisters. Then, within each of the forty-nine locations defined by ring and section can be found the memorable images themselves. For example:

> *The work is of wood, marked with many images, and full of little boxes; there are various orders and grades in it. He gives a place to each individual figure and ornament, and he showed me such a mass of papers that, though I always heard that Cicero was the fountain of richest eloquence, scarcely would I have thought that one author could contain so much or that so many volumes could be pieced together out of his writings . . . He stammers badly and speaks Latin with difficulty, excusing himself with the pretext that through continually using his pen he has nearly lost the use of speech . . .*
>
> *He calls this theater of this by many names, saying now that it is built or constructed mind and soul, and now that it is a windowed one. He pretends that all things that the human mind can conceive and*

> *which we cannot see with the corporeal eye, after being collected together by diligent meditation may be expressed by certain corporeal signs in such a way that the beholder may at once perceive with his eyes everything that is otherwise hidden in the depths of the human mind. And it is because of this corporeal looking that he calls it a theater.*[8]

The modern reader can imagine Erasmus—that very embodiment of reason and the great skewerer of human folly—laughing at all of this absurd gobbledygook. But perhaps he did not. We are, at the time of Giordano Bruno, still decades before Francis Bacon publishes *The Advancement of Learning*, enshrines the scientific method, and kicks off the revolution that defines the next 450 years. Moreover, it took a very long time for empirical research to bring light to all of the dark corners of superstition and mysticism. As late as the eighteenth century, Isaac Newton, perhaps the greatest of all scientific minds, still secretly toyed with alchemy. So it is not surprising that here, in the middle of the "enlightened" Renaissance even the finest thinkers of the age might still have casually lumped magic and the dark arts in with the rest of the natural sciences.

And before we congratulate ourselves for having at last escaped this mumbo jumbo, one need only consider the influence of Bruno's intellectual descendant, Aleister Crowley; the occult religion of theosophy; and the psychological writings of Carl Jung on twentieth-century art and culture. In one of the most famous modern short stories, "The Aleph," Argentine writer Jorge Luis Borges posits that hidden in a cellar is the ultimate memory theater: a single point that contains all other points, so that in peering in one can see the universe from every possible angle.

Meanwhile, most of us have spent a lot of our lives in modern "memory theaters" whose pedigrees have been obscured by the centuries. For example, as Yates noted, the most famous theater of the seventeenth century, London's Globe, the home of many of Shakespeare's play premieres, appears to have been designed along the lines of Camillo's plan. How perfect: the world's greatest memory plays performed in the world's most famous memory theater! And that's only half of the story. Remember all of those odd little "figures and ornaments" in boxes in the alcoves of the memory theater? They appear to have been the model for the popular "cabinets of curiosities" in the centuries to come and that would be the precursor of the modern museum.

CATHEDRALS OF MEMORY

Camillo's memory theater simply electrified Italy. And from there the excitement spread to France, where King Francis I invited Camillo to visit him in Paris—and to bring along his theater. He did so, and only added to his celebrity by announcing that the king was the only other human being to whom he would impart the operation of the memory theater. Whether Francis I understood a word of what Camillo told him (the fact that he had a stutter suggests a more organic reason why Camillo had such a prodigious memory) is anyone's guess; but the king sure wasn't going to admit he didn't get it.

So, why this universal excitement over such an arcane enterprise? The answer is probably that Camillo's memory theater resonated with the Renaissance in a fundamental way that is largely lost to modern observers. That's because, more than we can imagine, the people of that world had lived almost daily, and for centuries, with their own kind of memory theaters: cathedrals and churches.

We marvel today at the *aesthetic* power of the paintings, frescoes, and stained-glass windows of the great medieval Gothic cathedrals such as Notre Dame and Chartres, and Renaissance masterpieces such as the Sistine Chapel, but to the men and women of the time, their power seems to have lain just as much in their *instructional* power. In a largely illiterate age, the act of memorizing and remembering key stories of the Bible, important verses, and acts of the liturgy (such as the stations of the cross) would have been a difficult task using only spoken words. But to sit in one of the vast cathedrals during Mass and look up to see those stories and actions poised above you, illuminated by the sun and painted in the most vivid colors . . . that would have made learning and memorization a whole lot easier.

And this mnemonic tradition wasn't only preserved in the great cathedrals. In the English country village of South Leigh, just a few miles from where *Bodley 764* rests in the Bodleian Library, stands the little Church of St. James the Great. It is almost as old as the Norman Conquest. In 1869, when the church became a separate parish, its first vicar decided to inaugurate the new era with a general restoration. When he removed the whitewash from the walls, he was astounded to discover a collection of murals—several unique (including a Last Judgment)—dating to the fourteenth and fifteenth centuries.

One of these paintings, of particular interest to this book, is found near the south doorway: *The Virgin and St. Michael* (see this book's frontispiece). Dating from the fifteenth century—and overlaying a similar, smaller painting from a century before—it shows the archangel, in medieval armor and with great outspread wings, holding a huge sword in one hand and balance scales in the other. Kneeling in one scale pan is a tiny, newly dead soul, a frightened-looking man who is naked and praying for redemption. In the other sits a devil. The latter is anxiously motioning to his counterparts on the ground below to hop up to join him and help tip the scales in hell's favor. One of these other devils blows a horn to call for further assistance.[9]

What is at stake in this soul-weighing is visible below: the open mouth of a giant monster, no doubt representing eternal damnation. Devils are already shoving doomed souls into its mouth; one of the devils is even armed with the classic trident flesh-fork.

All might seem lost for our dead soul but for the presence of the Virgin, who stands atop a crescent moon (as does Mexico's Virgin of Guadalupe) beside the soul's side of the scale. She has broken the strand of her rosary and is dropping the beads one by one into the dead soul's pan for added weight. At the moment she is winning, but it remains a very close call on the dead soul's fate.

To the illiterate medieval sinner sitting in the Church of St. James the Great, the message of this mural was as powerful as any given from the rostrum or in the unreadable pages of the Bible upon that rostrum: The wages of sin are death, and an eternal death of unbelievable horror awaits unless you turn to God, pray to the Holy Mother to intercede for you, and pray the rosary every chance you get—because your fate may balance on a single one of those prayers.

It is an image as complex, nuanced, and powerful as any devised by Marcus Tullius Cicero or Giulio Camillo. It serves as another kind of access key to the world of hidden knowledge—only in this case it is to the deeper meanings of the continent's dominant religious faith.

This, perhaps more than anything else, explains why the Camillo memory theater was met with such acclaim. As do technological innovations today, it represented an appealing breakthrough in terms of cost, size, and performance—now you could build an entire Gothic cathedral in a matter of months in your country home or castle. It was even (comparatively) portable. And since its use was, presumably, for the greater glory of God—hadn't Francis I placed his imprimatur upon it?—the Church was unofficially happy with it.

And that was only half the story. Modern historiography has pretty much erased (although not in textbooks) the myth of the Renaissance as being a radical turning point in human history, of the powers of light overcoming those of darkness. There was simply too much enlightened work going on in the late Middle Ages to see it as simply a low step before the leap into the Renaissance. The Middle Ages saw the rise of universities, scholasticism, the rediscovery and translation of classical works, invention, and even, in the cases of Ibn al-Haytham, Robert Grosseteste, and Roger Bacon, real modern scientific investigation. Friar Bacon, as much as anyone, by bringing science into the world of empiricism and the age of encyclopedias, set the stage for the early scientific researchers of the Renaissance such as Kepler and Galileo.

But if the "rebirth" myth of the Renaissance has been mostly debunked (it still represented a remarkable restoration of vitality and optimism to Europe after the nightmarish fourteenth century), one powerful new feature of that era has only been underscored by this new research. This was the rise of *humanism*—the shift away from filtering all knowledge and memory through the tenets of faith and theology and instead viewing them from the perspective of man and his new empirical tools. Scholars, scientists, and even theologians had long chafed at the restrictions placed on them by the institutions around them, even as they translated the much freer, more sophisticated, and more "natural" writings of their ancient predecessors.

We have seen this older, ecclesiastical view in works as diverse as the bestiaries and from the great scholastics like Aquinas. The Renaissance's new anthropocentric world view is best exemplified by Leonardo da Vinci's portrait of the *Vitruvian Man*, the man inside a perfect circle, itself a mnemonic for that universal genius's belief that man was the *cosmografia del minor mondo*—the physical embodiment of the workings of the universe. The rise of perspective in art was another example of this new view: The artist no longer looked at the world through the eyes of faith—painting figures at sizes that represented their ranking to God—but rather sought to represent what was actually seen by the human eye.

So, too, did this new humanistic viewpoint induce others to study the nature of society without worrying about paying much more than lip service to a higher power. Thus came Sir Thomas More's *Utopia,* Thomas Hobbes's *Leviathan*, Baldassare Castiglione's *Book of the Courtier,* and most durable of all, Niccolò Machiavelli's *The Prince.*

But nowhere did this new intellectual freedom have a greater influence than in the world of scientific research. It is hard today to imagine how liberating—and psychologically comforting—it must have felt to scientists and scholars of science of the age to at last be able to reconcile what they saw to what they were "supposed" to see. Hundreds of years of counterfactual evidence—about the natural world, about the cosmos, about history—had accumulated and stymied the search for objective truth. Now, like dross on metal, it could be stripped away to allow the real shape of reality to shine through.

The celebrated result was one of the greatest periods of scientific discovery and innovation in human history, from Gutenberg's commercialization of the printing press to the European discovery of the New World, from the microscope to the telescope, from Copernicus to Galileo.

With the "facts" of the past at last questionable, European scholars embarked on a second great round of translation, this time to see what the ancients *really* said, without the explicit (and self-imposed) censorship of the Church. This revisitation of the classics produced a few surprises, including a new appreciation of the Greek philosophers and growing doubts about Cicero's authorship of *Ad Herennium.*

The first had the effect of raising interest in democracy, in individual conscience, and in the challenge of creating a perfect state. The second ultimately proved a major blow to the Art of Memory: If the great Master Tully hadn't written this second book of memory, then all that was left of his writings on the subject were a few obscure paragraphs in *De Oratore.* So perhaps this Art of Memory business wasn't all it was cracked up to be; maybe it wasn't the key to re-creating the greatness of Rome. One can imagine this news came with considerable relief to the legions of scholars who had exhausted their brains trying to memorize entire books.

The death knell for the mass obsession with memory theaters and the Art of Memory came with the widespread availability of published books. As private libraries began to fill with volumes, the choice quickly became between spending weeks and months stuffing one book into one's brain . . . or simply owning that book, thumbing through it when needed, and then going on to other texts. The answer was obvious and became even more inevitable as indexing became more accurate and common. The disappearance of the Art of Memory, only recently the toast of European universities and palaces, was so complete that it would be four centuries before Frances Yates would rediscover it as if it were a lost civilization.

THE DANGEROUS MONK

But there was one more act to the story of the Art of Memory: that of Giordano Bruno. The Dominican order in which Bruno was a monk never gave up on Cicero's authorship of *Ad Herennium* (they might have been right), and thus it was still in place for Bruno to take it up as a tool in his larger quest. And that quest, which made Bruno one of the great heroes of the Renaissance, was to both free his own work in philosophy, science, and even religion, from all of the myths and false beliefs of the Catholic Church and at the same time, delve into the deep mysteries of occultism and the secret knowledge of the ancient world.

If this all seems contradictory to modern eyes, it is because Giordano Bruno was a deeply contradictory character—part genius and part fraud, part man of faith and part bomb-thrower against that faith. He heroically challenged the Holy See and its orthodoxy by publishing a series of controversial books, yet at the same time spent much of his fifty-two years on the run from angry hosts and papal authorities—from Venice to Padua to Lyon to Toulouse, often taking on teaching posts along the way.

Arriving in Paris in 1581, Bruno embarked on a series of thirty theological lectures that made him famous—not just for the brilliance of the lectures themselves but for the prodigious memory he showed in giving them. Eventually the acclaim reached King Henry III, who, like his grandfather before him, invited the reigning monarch of memory to his court. Wrote Bruno about the experience:

> *I got me such a name that King Henry III summoned me one day to discover from me if the memory which I possessed was natural or acquired by magic art. I satisfied him that it did not come from sorcery but from organized knowledge; and, following this, I got a book on memory printed, entitled "The Shadows of Ideas," which I dedicated*

to His Majesty. Forthwith he gave me an Extraordinary Lectureship with a salary.[10]

This quote captures much about Bruno—his brilliance, his confidence, his fundamental intellect . . . and his general contempt for most other human beings. Next, armed with a letter of introduction from the French king, Bruno moved on to England. There he wrote some of his most important works, hung around with the Hermetic Circle there under John Dee, and again managed to create enemies with his sarcastic attitudes and controversial beliefs (such as the Earth revolving around the sun).

He returned to France, but after writing some controversial works against Aristotelian science, he moved on to Germany, where he spent the next five years. There, incredibly, given the war between Protestantism and the Catholic Church, Bruno still managed to get himself excommunicated by the Lutherans.

In 1592, having received an invitation to return to Venice and tutor a wealthy patrician, and apparently believing the Italian Inquisition had lost its teeth, he accepted. Within two months, when he announced he was leaving, Bruno had so angered his wealthy client that the man denounced him to the papal authorities.

For the next seven years, Bruno was imprisoned in Rome. His inquisitor, Cardinal Bellarmine, would gain notoriety for assuming the same duty twenty years later with Galileo. This connection, combined with some of the crimes for which Bruno was charged—including Copernican heliocentrism and the aforementioned belief in "a plurality of worlds and their eternity"—is what brought those crowds to his statue in the Campo de' Fiori in 2000 and has made him a patron saint of maverick thinkers.

But Giordano Bruno was less a martyr to science and more a martyr to religious orthodoxy—a much more common, but also much less romantic, legacy. For example, while it is appealing to think of Bruno as running afoul of the Inquisition for his support, (like Galileo) of Copernicus, the fact is that at the time of the trial the Roman Catholic Church had established no official position on an earth-centered cosmology. What the Church challenged was Bruno's belief (which we now suspect to be accurate) in an infinite universe of infinite worlds of infinite possibilities—including, by implication, infinite redemptions by an infinite number of Christs.

Indeed, they were very different trials, with very different results. Galileo, a scientist *and* a believer, was tried for being "vehemently suspect of heresy" for his astronomical writings and forced to renounce his publications and live under house arrest. Bruno was a cleric and a sworn member of the Church and he was properly accused of multiple counts of heresy, including expressing doubt in print about Christ's divinity (he called him a "magician"), the Holy Mother's virginity, the Trinity, the Mass, Transubstantiation, and the Incarnation—in other words, the absolute bedrock beliefs of Christianity itself. It is no wonder the Lutherans excommunicated him as well. This was true heresy. And while they were at it, the Inquisition threw in accusations of magic and divination. And the list of charges didn't include some of Bruno's more unsavory beliefs, such as a belief that the different human races descended from different sources, and thus weren't the same species.

Bruno didn't help his case by first refusing to make a full recantation and then by circumventing Bellarmine by taking his case (with only a partial recantation) to Pope Clement VIII. Once again, Bruno had managed to infuriate everyone with his contempt for both the people in authority and institutional protocol. And the partial-recantation proposal was, as usual, too clever by half. Rather than be moved by Bruno's request, Clement VIII asked for the death penalty. It was as he was being read his sentence and realizing that there would be no escape this time that Bruno reportedly lunged at his judges and made his famous remark.

All that was left was the stake and bonfire.

As the decades passed and his books retreated into dusty libraries, Bruno's beliefs were disproven and superseded, and his memory arts lost. Giordano Bruno became a curious, but minor, figure of the Renaissance. But he wasn't entirely forgotten; rather, he was transformed into a kind of mystical cousin of Galileo, who alone had died in defense of modern science. Copernicus had published only after his death; Galileo had been contrite before the Inquisition; but Bruno had not only chosen death but thrown his conviction back into the face of his accusers.

It was this Giordano Bruno who was remembered, and the man who was honored in 1889 with the evocative statue in the Campo de' Fiori—the result of an international commission that included author Victor

Hugo, playwright Henrik Ibsen, and social biologist Herbert Spencer. Ironically, it was this mythical Bruno who survived into the middle twentieth century to be rediscovered and restored in full by Frances Yates.

If the Art of Memory didn't survive the execution of Giordano Bruno, examples of prodigious acts of memory have continued to appear and be celebrated as superhuman achievements, right up to the present. There would always be prodigies of memory; men and women who could not only hold vast libraries of information in their brains but could also seemingly access any item in their memory at will, and with astonishing speed. In the eighteenth century they performed before kings and queens, in the nineteenth onstage before amazed crowds, and in the twentieth on television game shows. And though they might achieve these feats through some systematic and specific techniques for memorization, never again would memory be seen as an art form, available to everyone, and perfected by the best and brightest in society.

Rather, the usual assumption was that most of these prodigies had in fact been born with uniquely configured brains—what would eventually be medically classified as "Eidetic memory"—that seemed to feature a different interaction between the small quantities of usable short-term (what is now called "working") memory and the immense, but much less accessible, long-term memory. It was also noted that this enhanced talent often came at considerable cost—autism, lack of creativity, even brain damage. And so while these memory prodigies were often the subject of awe for their powers, that awe was rarely accompanied by respect for their character or for any hard work in earning their gifts.

Still, in the centuries that followed the death of the Art of Memory, there were a few occasions where, often under duress, human beings not only exhibited prodigious memory but did so at the service of something even greater.

Perhaps the greatest of these took place just sixty years after the death of Bruno. The British poet John Milton, blind and reduced to poverty, composed and dictated to his daughters—day after day, mentally moving through the text to make revisions and edits—three of the greatest poems in the English language: *Paradise Lost, Paradise Regained*, and *Samson Agonistes.* He may have written these poems to, in his words, "Justify the ways of God to man," but readers couldn't help but see Milton himself in the blinded Samson, "Eyeless in Gaza, at the mill with slaves." It stands as one of the most remarkable feats of memory and creativity ever accomplished.

Two hundred fifty years later, just a few years after the raising of Bruno's statue, a young Parisian, Marcel Proust, would embark on a work of literary memory almost the equal of Milton's. Retreating to a cork-lined room, he would write *À la recherche du temps perdu* (the title taken from a Shakespearean sonnet: "in remembrance of things past"), a book that begins a half-century memory with the taste of a madeleine cookie—and famously ends, three thousand pages later, in the present, at a party in which the characters of the book appear, aged and doddering as if on stilts built from lifetimes of memories. The author/narrator decides at that moment to write the book we have just read. It is an act of memory restoration, and artistic transformation, without equal.

LOST MEMORY

The execution of Giordano Bruno may have signaled the end of the Art of Memory, but by then, the technique was almost obsolete anyway. It had proven too difficult, too much in the hands of strange characters with mystical and heretical beliefs. And too many people suspected that its successful implementation depended as much upon the innate abilities of its users—as well as a whole lot of traditional memorization by repetition—as it did upon any special tricks or dark arts.

But the real credit goes to Johannes Gutenberg. And to paper. In the war against the Art of Memory for the future of "synthetic"—what we'll now call, modern phrasing superseding an old definition, "artificial"—memory, paper only emerged victorious because of printing. But printing also won because of paper.

By the fourteenth century, paper was a thousand-year-old technology in China, and production there was measured in the millions of sheets per year. Paper mills had been in use in Persia for seven hundred years, and in Europe—in Spain, at least—for four centuries.

Printing was even older. The Mesopotamians had pressed cylinder seals into their clay tablets as early 3500 B.C. Woodblock printing first appeared in China in the third century A.D. and in Egypt a century later. Moveable type, the great breakthrough that enabled the creation of different pages from a single print bed, is credited to Chinese printer Bi Sheng in A.D. 1040. His type, originally made of clay, was upgraded to carved wood and then, in Korea in 1230, to metal. In the meantime, the printing of imagery had evolved to a very high standard with the development of wood printing, lithography, and copper engraving and, in the hands of masters like the artists of China's Sung dynasty and Germany's Albrecht Dürer, into some of mankind's greatest works of art.

The stage was now set in the mid-fifteenth century, in an almost unique historic moment, for what modern technologists might call a *double, mutually beneficial tipping point.*

Johannes Gutenberg was, by most accounts, a moderately successful metalworker who had eked out a living as a goldsmith and gem polisher. But after one particularly risky venture—selling polished metal mirrors to pilgrims to capture the holy light emitted by religious relics—Gutenberg found himself not only broke but being pressed by his investors. That's when, in the classic misdirection move of a born entrepreneur, Gutenberg announced to those creditors that he had an even greater invention waiting in the wings.

It took him ten years, from 1440 to 1450, to finally roll out a full working version of this invention—the *printing press.* Portentously, one of the first print runs from this press was papal indulgences—the notorious documents by which people could buy their way into heaven.

In 1455, even as he was printing those indulgences, Gutenberg set to work on the project that marked one of history's greatest turning points: the publication of the forty-two-line-per-page "Gutenberg" Bible. It was the first truly printed book edition—and today the surviving copies are the most valuable books in the world.

What often goes unnoticed about these first 180 Gutenberg Bibles is that the production was divided between versions on paper and on vellum. This mix of media signals a separate, but also important, moment of transition. It suggests that, with almost perfect timing and symbolism, the two adoption curves of paper and parchment have just crossed, with the latter trending toward oblivion and the former toward universal adoption.

In retrospect, if not at the time, this technological turnover seemed inevitable. Parchment had many wonderful qualities, not least its durability. But for every short stack of parchment sheets you had to slaughter a small herd of cows or sheep: and for vellum, calves. That would be an efficient use of animal flesh left over from butchering. However, that same flesh could also be converted into equally valuable leather. A modern economist would say that there is a high "opportunity cost" for choosing to turn animal flesh into parchment rather than leather.

But even if leather had never been invented, there is still a problem: Parchment production becomes inefficient very quickly should demand for this product exceed the demand for meat. Then you have a choice: Accept skyrocketing prices for parchment or start slaughtering your breeding herd (and the new generation, now calves) for short-term gains or put in new pasture for long-term reward. Whatever your choice, it is going to be costly, because (to use modern business terminology) parchment doesn't effectively *scale.*

Now consider paper. Essentially made of weeds, old rags, and common chemicals, it can be formed in almost any size, including rolled sheets hundreds of yards long, and easily cut to order. As such it is one of the most scalable physical products ever devised. It may be more fragile than parchment, but that problem is easily solved by binding the entire manuscript in leather (rather than making the manuscript itself entirely of skin).

In other words, paper was destined to eventually triumph. But the printing press turned that "eventually" into "now."

The advent of paper and printing also had a profound effect on literacy in Europe—though not just in the way it is usually taught. Unquestionably, the appearance of inexpensive printed textbooks and manuals transformed and expanded education. But there was another, less recognized, effect: It taught already-literate people, for the first time, to *write.*

WRITING ALOUD

We often think of people of the Middle Ages as being almost universally illiterate, both nobles and serfs, with the exception of a few ecclesiastical scholars. The truth was much different, says historian Michael Clanchy. As early as 1170, British king Henry II, in establishing a government bureaucracy, instituted a formal inquest into the financial dealings of his sheriffs and other officials:

> *Likewise inquiry is to be made concerning the archbishops, bishops, abbots, earls, barons, vavasors, knights, citizens, burgesses and their*

> *stewards and ministers as to what and how much they have received throughout their lands from every one of their hundreds, every one of their villages and every one of their men—whether by judicial process or without judgment—and they are to write down individually all these exactions and their causes and occasions.*[11]

The use of the phrase "write down" suggests a level of general literacy among individuals of authority in England far higher (and far earlier) than generally assumed. But as Clanchy reminds us, this isn't exactly the case.[12] As the quote above implies, the ability to *read* was much more common (and in demand) at this early date than we usually assume. But because we live in a world where reading and writing are considered two sides of the same coin, we take for granted that this has always been the case. But in the late Middle Ages it was not the case—and for a very interesting reason:

> *Because it was more difficult to write with a quill on parchment than it is with a modern ballpoint pen on paper, writing was considered a special skill in the Middle Ages which was not automatically coupled with the ability to read.*[13]

In other words, when we speak of literacy in the three centuries before Gutenberg, the proper phrase isn't "reading and writing" but "reading and *dictation*." An uncounted number of people in authority in Europe, the Middle East, China, and Japan were able to read but were barely capable of writing much more than their own names. The result was the creation of a new class of professionals—*scribes*—who filled the gap between the nobleman's words and the scratching of letters on parchment. An interesting secondary effect of this professionalization of writing was the development of a new, more efficient type of writing, *cursive*, that dominates handwriting to this day.

Then along comes paper, with its smooth surface and ever-cheaper price. And with it, slowly but inexorably, Europe and much of the rest of the world finally learned to write. Once again, the effect of this shift on mankind's memory due to the resulting flood of personal letters, diaries, and notes is incalculable.

A MATTER OF SCALE

Paper made the medium of memory scalable for the first time; printing made the *content* of that medium scalable as well. Now an author could write one manuscript and it could be printed into one hundred, one thousand, or even one million copies—each at a very small per-unit cost rather than follow the old handwritten paradigm in which each new copy cost as much as the first.

The result was a revolution in the *business* of books. Booksellers quickly found that there was more money to be made in large sales of low-cost printed books than in high-priced single copies of handwritten books. Printers, in turn, had a financial incentive to improve their processes through innovation. And authors found that their writing could earn them not only a reputation but also an income. Once again, this didn't occur overnight, but in time it would lead to an explosion in book production that would at last put texts in the hands of even the lower classes, fill uncounted numbers of private libraries, provoke the creation of public libraries . . . and ultimately lead to the turbulent, thrilling, and corrupt world of authors and booksellers, such as existed in Paris in the nineteenth century, immortalized by Balzac in *Lost Illusions*.

The one-two punch of scalability made possible by paper and printing—and the revolutionary new business model it established—in turn began to shift power across Europe and Asia, and eventually the New World. Martin Luther may have handwritten the original 95 Theses he nailed to the door of the Castle Church in Wittenberg in 1513, but it was in printed form that this document was disseminated throughout Germany and sparked the Protestant Reformation.

Indeed, the Protestant Reformation at its beginning can be seen as the battle between Gutenberg's first printed product (the indulgences) and the second (posters and books). Protestant publications, produced by the hundreds of thousands, proselytized millions across the continent faster than the slow-moving old Catholic Church could respond. Not long afterward, these Protestant victories were made permanent with the appearance of new everyday-language ("vulgate") versions of the Bible, including a German version by Luther and, most famously, the unfinished Tyndale English Bible (1525–1530), which was the unlikely antecedent of one of the greatest achievements of the Late Renaissance, the

King James Bible (1611). "Unlikely" because Tyndale was executed for heresy in 1535, for creating that Bible. Reportedly, as he died, Tyndale shouted, "Lord, open the King of England's eyes!"[14]

His prayer was answered. And when Bibles began showing up not only on the altars of churches but in private homes, the monopoly of Roman Catholicism over everyday life in Europe and the New World was broken forever. But so, too, was the control of the new Protestant denominations over their own adherents.

Finally, as every businessperson knows, the search for profits inevitably leads to determining what the customer wants and what he or she will pay for it. And what printers and booksellers discovered then, and what remains true today, is that there is always a demand for compilations, collections, and surveys—both for entertainment and self-instruction.

This demand had underwritten the creation of books like *Bodley 764* in the Middle Ages, and now, in the Renaissance, the demand was many times greater. And the ability to meet that demand, thanks to paper and printing, was greater as well. Moreover, this demand also dovetailed neatly with two intellectual trends that had only grown over the past centuries: scholasticism, and the ongoing effort to accurately organize all of human knowledge both past and present.

In the age of Aquinas, such works had been both written by hand and in very small volume—and they had also been filtered through the perspective of the Roman Catholic Church. In the Renaissance, the process was mechanical and produced in large volume. In addition, in an increasingly humanistic world, almost all of the earlier works and translations were under suspicion for inaccuracy and bias. Add to this an increasingly wealthy and literate audience hungry for access to learning and memory, and all of the ingredients were in place for a third, multigenerational boom in encyclopedias, books of natural history, atlases, dictionaries, glosses of religious texts, and instruction manuals.

The first of this new generation of encyclopedic texts, ranging from treatises on painting to books of engineering, began to appear in the sixteenth and seventeenth centuries. The most influential of these early works were bilingual dictionaries designed to assist travelers and diplomats.

Then in 1592, Richard Mulcaster, a teacher, compiled the *Elementarie*, a lexicon of eight thousand English terms to help his students remember the spelling of difficult words. Twelve years later, another teacher—Robert Cawdrey—published what is considered the first true English dictionary, *A Table Alphabeticall.* In France, the effort would take another hundred years, but the result—*Le Dictionnaire de l'Académie française*—would prove definitive right from the start. Similar efforts took place throughout Europe during this era. Even as far away as Japan, Jesuit missionaries produced in 1603 the *Nippo Jisho*—a bilingual Dutch-Japanese dictionary. Japan had already known dictionaries, derived from the Chinese, for hundreds of years, but the *Nippo Jisho* had the unexpected secondary effect of formalizing modern Japanese.

But the real explosion in what we now call reference books began in the eighteenth century, producing a number of works that, dozens of editions later, are still with us today. Among dictionaries, the most famous creation was *A Dictionary of the English Language* (1755), a work as brilliant and eccentric as its sole author, Samuel Johnson. Beginning at the end of the eighteenth century and working for more than a quarter-century, Noah Webster created *An American Dictionary of the English Language*, which permanently established the differences between the two dialects. All of this work culminated in the final decades of the nineteenth century with the publication of two great multivolume dictionaries, both of them archiving the memories of their respective languages. They remain the standard today: *The Oxford English Dictionary* and the *Larousse Grand Dictionnaire Universel.*

Encyclopedias followed a similar trajectory. As already noted, encyclopedias, both general and subject-oriented, had been created in earnest during the high Middle Ages across the globe, most notably the gigantic fifteenth-century *Yongle Encyclopedia* of China's Ming dynasty. During the Renaissance, feeding a hungry public armed with enough money to purchase a single all-encompassing book, encyclopedias were everywhere. But, as with dictionaries, encyclopedias took on their modern form during the Enlightenment—the last time in history (it is said) that a person might presume to know everything.

The three great encyclopedias that marked this era were products of creators who might be described as neo-scholastics. The first was Ephraim Chambers's British *Cyclopedia*, published in 1728 and considered the first modern encyclopedia. Third was the 1796–1808 German *Conversations-Lexikon mit vorzüglicher Rücksicht auf die gegenwärtigen Zeiten* (Encyclopaedia with special regard for the present time), known today as the *Brockhaus Enzyklopadie.*

But the middle encyclopedia is today the most celebrated, the French *Encyclopédie*, edited by Jean le Rond d'Alembert and (famously) Denis Diderot in seventeen volumes, from 1751 to 1765. The *Encylopédie* is justly celebrated for its quality and its influence on French culture before the Revolution. It is also celebrated because Diderot and his fellow "encyclopedists" represented not just a publication team but a new and objective view toward human knowledge—in Diderot's words, "to collect all the knowledge that now lies scattered over the face of the earth, to make known its general structure to the men among [whom] we live, and to transmit it to those who will come after us."[15]

This new perspective, of making people not just smarter but *better*, would reach its zenith at the dawn of the nineteenth century with the work that would dominate reference libraries for the next two centuries: the *Encyclopaedia Britannica.*

OUT OF MIND

The zeal for classification during the High Middle Ages and early Renaissance didn't just infect scholars, occultists, and ecclesiastics but also the first modern scientists. Speaking across the ages, the recovered Aristotle had called upon all scientists to look beyond what ought to be true to what really *was* true—that is, to observe the natural world.

The protoscientists of the Middle Ages, such as Roger Bacon and al-Haythem, took this process one step further from the pure deduction of Aristotelian science to the much more ambitious *induction* of the scientific method—a process codified by Francis Bacon and enshrined in institutions such as the Royal Society of London, the Academie Française, the Italian Accademia, and the German Academy—all founded in the seventeenth century and all dedicated to systematizing and improving scientific research.

The result was not only an exciting new era of scientific discovery but, behind all of those great discoveries, mountains of observational data and volumes of lesser discoveries. As these caches of information grew they posed their own problems, slowing new research and inhibiting dissemination. Needless to say, the field was ripe for organization—in particular, for classification schemes that not only put all of this knowledge in its place, free of superstition and prejudice, but also added to the overall understanding.

The names of the scientists who first took on the challenge of compiling, organizing, and classifying the accumulated knowledge in their disciplines are often honored today as the "pioneers" of their fields. Among the first were the physician Andreas Vesalius with his groundbreaking work (1543) on human anatomy and the physicist William Gilbert on electricity and magnetism (1600). Later classifiers included Carl Linnaeus, the father of modern taxonomy (1735) and the man whose *Systema Naturae* gave a Latin designation to all species, living and extinct; Boyle and Lavoisier on chemistry; Dalton on atomic theory; Mendeleev with the periodic table of elements; Newton on classical mechanics; Charles Lyell on geology; and Alexander von Humboldt on geography. Even the pictorial works of John James Audubon are part of this tradition.

By the end of the eighteenth century, almost every scientific discipline had been given structure, oversight, and a system of classification. And these seminal scientific works—along with the encyclopedias, the natural histories, and ethnology picture books; the atlases, dictionaries, and biographies; the compiled speeches and national and military histories—had in their own unexpected way managed to accomplish what Bruno, Camillo, and even Cicero had failed to do: They had put all of human memory within reach of the average person, no matter what that person's natural aptitude.

The Art of Memory had struggled to find a biological solution to the challenge of retaining massive amounts of memory inside the human brain. Its solution was to define a new kind of information retention, and come up with extremely complicated tools to supercharge it.

The High Middle Ages and the Renaissance, venerating the ancients, at first tried to revive the lost memory arts. But in the end, and facing a far greater memory challenge, it found instead a technological solution: printed books. This new type of memory, found outside the human body, was not as portable, but it was much more scalable and more universally useful. Just as important, it was *transferrable*.

And if, in the world of artificial memory, there were no bloody statues in elegantly rendered memory theaters, there were tables of contents and indexes, footnotes and bibliographies.

It had taken 1800 years, since Marcus Tullius Cicero first walked through a temple to memorize his first speech, for mankind to preserve the memory of everything it had known and learned. But that day had at last arrived.

Meanwhile, in the intervening centuries, mankind, as a result of this new knowledge, had also begun to populate its world with machines. These machines, from simple household tools to mammoth constructions, increasingly filled every corner of society. And for all of the benefits and bounty they delivered, their operation and maintenance consumed ever-greater portions of everyday life.

If human memory could be removed from the human brain and placed into books, charts, and tables . . . might it be possible to put this same artificial memory into machines as well?

7
Patterns in the Carpet
Memory as Instruction

The most ecstatically received new invention in mid-eighteenth-century France, the iPod of the reign of Louis XV, was a machine that magically produced . . . duck poop.

It was called the *Canard digérateur,* or the Digesting Duck, and it was made of gilded copper feathers, papier mâché, rubber hose, wheels, gears, and armatures—four hundred moving parts, all artfully constructed to resemble a real duck. The feathers were even perforated to make the duck "transparent" so that viewers could look inside its body cavity to confirm the presence of mechanical internal organs congruent to those found in a living counterpart.

Thanks to a system of cams and nearly thirty levers, the Digesting Duck could dip its bill into water, make realistic gurgling sounds as it drank, stand, lie down, stretch and curve its neck, and even move its highly articulated wings, tail, and larger feathers.[1]

What made the Digesting Duck a sensation was one other singular capability: If you fed it pieces of corn or grain and waited a few moments—apparently how long it takes for mechanical digestion to take place—it would defecate (from the right place) a neat little duck turd. Not the messy duck droppings of real life, but real enough—and more appropriate to land on a Versailles tabletop.

But what made the Digesting Duck more than just a brief sensation and something of epic importance was that it opened the door to—and set the genius behind its invention on the path of—two of the greatest technological and social transformations in human history.

That's a lot to make from a little duck poop.

MECHANICALLY THOUGHTLESS

As we saw in the Chapter 6, the explosion of accumulated knowledge and information created by the Renaissance had inspired a desperate need to organize these mountains of data into some intuitive and accessible form. But those new dictionaries, encyclopedias, manuals, and specialized libraries only brought into stark relief another problem: All of that newly cataloged and codified competence, while doing wonders for the training and education of human beings, did nothing for the machines and devices they increasingly devoted their days to operating.

This wasn't a real concern in the sixteenth and early seventeenth centuries. The application of the new indexing tools and stored expertise in support of more universal education had powered miracles of productivity improvements, the tools of global exploration, and the rise of the professions.

But for all of these improvements, the machines of this new "enlightened" world were still essentially extensions of manual labor. The tools might get more precise and powerful, but ultimately their every action had to be managed by a human operator. As such, while they might liberate their operators from actual physical danger, they were *not* labor-saving devices; on the contrary, the greater demand for manufactured goods by the growing middle class only meant more hours of skilled people manning machines that couldn't be handed over to an apprentice or left running on their own.

What was needed now was a way to make machine production scale up without dragging their poor skilled operators along with them. Such a solution—which would require the machines to become more *autonomous*—would not only have the cost-cutting advantage of (somewhat) liberating their operators but would also free the machines themselves to reach levels of speed, precision, and consistency impossible with fallible human interference.

But how to do it? The older, purely mechanical machines, such as looms and forges, simply stopped functioning once the operator walked away. The new water and spring-powered machines, from mills to clocks, were impressive but monomaniacal: They did one thing very well, but unless a human operator intervened, they'd continue doing what they did until they wound down to exhausted stasis or froze up from unrelieved friction.

But how did you make such machines remember how to perform? How did you make them take multiple steps in a task without having a person there to intervene and control the transitions? And most of all, how did you transfer all of the expertise that had just been collected, indexed, and published *into* the "minds" of these machines so they could not only use it, but improve upon it over time?

The answer came, as it often does, from the most unlikely quarter. In this case, the expensive toys of the idle rich.

LIVING STATUES

The Digesting Duck was, in a word that has taken on a somewhat different meaning in recent years, an *automaton*. Even by the time it was built by a watchmaker's son, Jacques de Vaucanson, in 1739, automatons already had a history that stretched back a couple thousand years.

We know, for example, thanks to discoveries such as the recovery in 1900 of the Antikythera mechanism—an amazing mechanical computer for calculating astronomical positions—that even classical Greeks could build sophisticated machines.[2] It apparently didn't take long to use those geared "engines" to empower sculptures to move and imitate nature. As early as the fourth century B.C., the great lyric poet Pindar would write of the automaton makers on the Island of Rhodes that:

The animated figures stand
Adorning every public street
And seem to breathe in stone, or
move their marble feet.[3]

As we saw in Chapter 4, the ancient world could be a stranger place than we imagined—and the image of streets lined with automatons certainly fits that notion (and the representation of a stunning figure, in this resonant location, obviously draws parallels to the Colossus).

And Rhodes wasn't alone: Across the Adriatic, in the Corinthian colony of Syracuse, similar mechanisms may also have been invented by the most celebrated ancient engineer of all, Archimedes. We do have a record that suggests he had constructed a kind of orrery, or planetarium, that used gears to show the orbits of planets and stars. From there it would have been a short jump to mechanical creatures.

Indeed, in the second century B.C., it was common for wealthy children in Hellenistic Greece to play with mechanical toys. Hero of Alexandria, often considered the greatest experimental scientist of the ancient world—the man who invented everything from the syringe to the windmill to the steam engine to the vending machine—famously put on a ten-minute-long play performed entirely by mechanical devices.[4]

Frankly, the more you look around the ancient world, the more automatons seem like the norm rather than the exception. Some of the examples are disturbing, such as the notorious brazen bull of Sicily, appearing in various versions for hundreds of years, which used the screams of victims being roasted to death inside a bronze bull to create the bull's "voice." Other examples are also remarkable—not least the Chinese account, reputedly from earlier than 1000 B.C., of an engineer named Yan Shi presenting to King Mu of Chou a life-size human automaton:

> *The king stared at the figure in astonishment. It walked with rapid strides, moving its head up and down, so that anyone would have taken it for a live human being. The artificer touched its chin, and it began singing, perfectly in tune. He touched its hand, and it began posturing, keeping perfect time . . . As the performance was drawing to an end, the robot winked its eye and made advances to the ladies in attendance, whereupon the king became incensed and would have had Yen Shih [Yan Shi] executed on the spot had not the latter, in mortal fear, instantly taken the robot to pieces to let him see what it really was. And, indeed, it turned out to be only a construction of leather, wood, glue and lacquer, variously coloured white, black, red and blue.*[5]

Being a king, Mu decided to disembowel the automaton. He reached into the creature's body cavity and started yanking out one internal organ after another. Incredibly:

> *The king tried the effect of taking away the heart, and found that the mouth could no longer speak; he took away the liver and the eyes could no longer see; he took away the kidneys and the legs lost their power of locomotion. The king was delighted.*[6]

No kidding. Ancient Chinese history also records a flying-bird automaton, invented by a certain Kungshu Phan, which reportedly flew for three continuous days.[7] Meanwhile, Jewish tradition holds that Solomon himself designed a throne that, when he ascended (in other words, sat down on) it, mechanical animals around the frame hailed him as king, a mechanical eagle lowered a crown onto his head, and a dove presented him with a Torah scroll.

Religion, violence, entertainment, toys—basically anything that inspired awe in the audience—seem to have been an occasion for automatons in the ancient world. It is a theme that continues into the present with the latest generations of humanlike robots coming out of Japan. Automatons also seemed to have enjoyed periods of great popularity (such as in ninth-century Persia and Renaissance Europe) and during other ages (the medieval world, for example) only to be nearly forgotten.

A GHOSTLY ENCOUNTER

Seventeenth-century France was just such a period of obsession with automatons, many of them prototypes of devices we still see today: birds that pop up and sing, animated dolls, and music boxes. Not coincidentally, at the same time, next door in Bavaria, clockmakers were inventing the most recognizable and popular automaton of all: the cuckoo clock. The young Sun King, Louis XIV, was entranced by a spectacular mechanical coach and horses complete with driver, footman, page, and an elegant lady passenger.

But something had changed. And it was due to the great philosopher and mathematician René Descartes. In detaching the mind from the body in the bath one day in 1641—the *Meditations on Philosophy*, the transitional moment between ancient and modern philosophy—Descartes not only sent philosophy on a centuries-long quest to reconnect them both but also raised two questions that have largely defined modern philosophical inquiry:

1. How does the mind work?
2. What constitutes "life"?[8]

It was the latter that initially got the most attention. If human beings (and perhaps other living things) were, in fact, "ghosts in machines," how did those machines function? Not surprisingly, everyone immediately drew parallels to those machines that behaved most like they were alive: automatons.

Descartes underscored this with some notions of his own—in particular, his often-stated belief that animals themselves were merely machines in which tissues, muscles, and nerves were simply God's more perfect (albeit messier) versions of pulleys, wires, and gears. Pursuing this still further, Descartes argued that since animals (unlike humans) lacked the divine spark of consciousness—the cogito in his *cogito ergo sum*—they couldn't really think. In other words, animals were merely wonderfully complicated automatons. And this in turn convinced Descartes that animals couldn't really feel pain, a notion that persists, against all evidence, in some populations to this day.

This image of animals as ghostless machines produced two trains of thought, one ghastly and one that is the subject of this book. On the nasty side, the belief that animals were incapable of experiencing pain led to the popularization (which still survives in some high-school biology classrooms) of *vivisection*—the science of dissecting living animals—with generations of scientists and doctors managing to overlook the shrieking and flailing of the creatures on which they were slicing away.

The second, and blessedly the more influential, pathway from Descartes's notion of the body as machine led not only to automatons but also to a fundamentally new way of thinking about these devices. As Stanford history professor Jessica Riskin describes it, "The designers now strove, not only to mimic the outward manifestations of life, but also to follow as closely as possible the mechanisms that produced these manifestations."[9]

In other words, it was no longer enough for automaton makers to build clever mechanisms that exhibited lifelike behavior; now automatons had to *be* lifelike. In Professor Riskin's words, "the design of automata became increasingly a matter, not just of representation, but of simulation."[10]

After a quarter-millennium of research into cells, membranes, nucleic acids, oxidation-reduction, hormones and glands, reproduction, digestion, and a thousand other features and processes of living organisms, we now know that simulating living organisms is infinitely more complicated than any eighteenth-century scientist or watchmaker ever imagined. Indeed, the automaton makers ran into this problem almost from the beginning . . . which is why they often cheated.

Consider again that story about King Mu. It's pretty obvious that the human automaton operated using the usual components, but ones that were disguised to look like real human organs. That's why the automaton lost its voice when its "heart" was removed.

By the eighteenth century, and amid the growing pressure to build devices that duplicated living things, automaton makers almost had no choice but to cheat. Take Jacques de Vaucanson, arguably the greatest automaton maker of all. In 1737, he wowed the French royal court with "The Flute Player," usually considered the world's first true biomechanical automaton. But when it came time to top that wonder with an even greater device—his masterpiece—de Vaucanson found he had hit a technological wall.

And that's why the Digesting Duck, the wonder of its era, turned out to be a fraud.

DUCK CONFIT

In fact, the duck did perform most of the actions claimed of it. It just didn't do the one for which it was most famous: pooping. It took thirty years for the fraud to be uncovered. In truth, the meal fed into the duck's bill fell down its hollow throat onto a tray, where it remained. Meanwhile, the hindquarters of the duck were prefilled with fake duck turds (crumbs pressed together and dyed green) and a mechanism pushed them through an "intestine" (one of the very first identified uses of rubber hose) out through the spot on the duck where the sun didn't shine.

The reader might be asking at this point why more perceptive observers of the era didn't detect the deception merely by the fact that in the few seconds between the feed going into one end of the duck and the feces coming out the other that there simply wasn't enough time for digestion to take place. After all, even the nobility had likely spent a bit of their lives in relatively close proximity to farm animals.

The precise answer is unknowable at this distant remove, but it probably had to do with a combination of context, credulity, and complexity. Even today, experiencing a major technological breakthrough for the first time is always an event surrounded by a kind of magical awe, where the apparent impossibility of what you are seeing frees you to overlook what later seem to be obvious flaws. Looking at those old automatons and their replicas today, it's hard to imagine how anybody could have been fooled into believing these creaky, jerky machines were alive. And yet I can remember as a boy in the early 1960s sharing in the audible gasp of a crowd at Disneyland as they watched the "animatronic" Abraham Lincoln for the first time.

For an even more contemporary example, consider how, even as we were dazzled by our new ability to download MP3 music files by the thousands into a tiny, handheld device, we barely noticed that the sound quality of these files was vastly inferior to the LP records and cassette tapes they would soon replace.

In other words, once the poop hit the tabletop—or, more precisely, a silver tray—nobody in the cheering crowd asked themselves why it had happened, and so quickly.

The Digesting Duck, with its winning combination of world-class construction, clever ruse, and earthy behavior, raised the bar on every clockmaker and automaton builder in the world. Its effect was so great that its image has regularly cropped up over the years all the way to the present—most famously in a Nathaniel Hawthorne short story ("The Artist of the Beautiful") and the Thomas Pynchon novel *Mason & Dixon.*

As de Vaucanson had taken the technology as far as it could go for the era (and, it seemed, further), the only way to top him in the automaton arms race was to extend the repertoire—that is, to make new devices that performed ever-more-complex and more far-ranging activities. After all, as any automaton owner knew, even the most elaborate, life-imitating device could only do a handful of predetermined actions over and over, while real living things were in a process of continuous and ever-adapting interaction with the outside world.

But if physical simulations were tough, imitating mind and thought was simply off the charts. Some builders just gave up and took the easy path of straight-out fraud. Some of these cons were so clever—such as Wolfgang von Kempelen's "The Turk," a purported chess-playing automaton that was in fact operated by a midget inside it—that they are still remembered (and occasionally imitated) to this day.

But others, still in thrall to the Cartesian paradigm, chose to keep trying to find a mechanical counterpart to the animal, if not human, brain. To do that, they turned to another, older automaton component: the pinned barrel. By the time of the Digesting Duck, the pinned barrel (or toothed cylinder) was already a century-old technology.

Probably everyone reading this book is familiar with this component, as it survives to this day in music boxes, regulator pocket watches, and, most famously, jack-in-the-boxes. It is essentially a metal cylinder out of which projects, at various predetermined locations, small spikes. Using a hand crank that is attached to a reducing gear, you rotate the cylinder slowly and with a fair amount of torque (thus the gearing), in close proximity to a collection of levers or, more commonly, a fine metal comb. In the case of the latter, as in a music box, the spikes hit the tuned teeth of the comb in the right order to create a melody. In the former, such as with an automaton, the spikes work the levers to produce the lifelike effects.

ONE THING AT A TIME

The pinned barrel seems prosaic today only because we live in the world it created. In the context of its day, it was an extraordinary invention. Why? Because for the first time it introduced time—or, more accurately, *timing* and *sequence*—to autonomous machines. Without the presence of a pinned barrel, all of the actions of an automaton would essentially happen all at once and continue until the power source ran down. But the rotation speed of the barrel set the pace of various actions, and the pins determined the order—as well as the start and stop times—of those actions.

And there was more. Already mentioned was the toy builders' growing need to expand the repertoire of their automatons—and here was a way to do it. Once they figured out how to make the barrels removable and interchangeable, they merely had to switch barrels with different pin arrangements (rather than rebuilding the entire device each time) to create a different performance.

This, too, may strike the reader of today as a minor matter, but in truth that little change represents one of the greatest intellectual breakthroughs. For the first time, mechanical memory, in the form of *instruction*, was detached from the machine itself. In more modern terms, it was at this moment that the paths of hardware and software diverged.

It can't be entirely a coincidence that almost exactly at this moment, the greatest philosopher in the world, Immanuel Kant, the heir to Descartes's mantle, was pacing the seven bridges of his hometown of Konigsberg, Prussia, and pondering the nature of human knowledge. Nor that he would choose to divide that knowledge into that with which we are born (a priori) and that which we learn (a posteriori). Descartes may have separated the mind from the body, the ghost from the machine, but Kant was determined to put some of operations of that machine—what we would call *firmware*—back into the functions of the mind.

But that's only half the story. Implicit in the operation of the pinned barrel is the possibility of *reversing* its operation. That is, taking an existing sequence of events and using it to determine the proper location of the pins on the barrel—thus making it possible to endlessly repeat even a complex activity or to capture an event for future study. That technology would find its way into every proper American parlor (and every improper American saloon) in the late nineteenth century as the player piano.

But the ultimate effect of this process reversal was far greater than that. If initiating action with the pinned barrel was the birth of instruction software, capturing information with that same barrel was the beginning of memory storage. Two hundred years hence, in the age of computers, this difference would define *read-only memory* (ROM) and *random-access memory* (RAM).

THE RULES UNROLLED

Remarkably, we still have one more, and even bigger, revolution to go. There is one simple act that, though largely unheralded, seems to consistently change history in a profound way. Simple as it may sound, it is the process of taking a successful new invention . . . and *flattening* it. That is, converting it from a three-dimensional to a two-dimensional form (or as close as you can get in the real world). We've already seen this with the printed book, and it happens again with the integrated circuit.

The single act of flattening the pinned barrel may have had a historic influence at least as great as the other two. To understand how this happened, we need to return to that genius designer, Jacques de Vaucanson.

From the perspective of the twenty-first century, de Vaucanson seems the very embodiment of the Enlightenment: worldly, multidisciplined, and infinitely curious. Yet at the same time he also seems more a figure from the future: a born entrepreneur, perpetually driven to pursue the next opportunity and outrun the competition.

Perhaps what drove him on was the familiar combination of enormous talent and a tough start. He had been born into the family of a master glove maker in Grenoble, France, and trained in a Jesuit school there. Unfortunately, de Vaucanson's father died when he was seven, and the family was left impoverished. It is said that even as a boy he planned to become a watchmaker and once, while waiting in an anteroom to make confession, had figured out the mechanism of a clock on the wall.

In time, de Vaucanson took orders with the ultra-rigorous Minims, an obscure order of Catholic monks that had a brief popularity.[11] Joining such a group was seen by de Vaucanson as the only way, in his poverty, that he could afford to continue his research. The Minims would prove important to de Vaucanson for an unlikely reason: They had been deeply influenced sixty years before by one of their members, a Father Mersenne, who was a dedicated scientist and friend of Descartes's—hence de Vaucanson's internalization of Cartesian dualism.[12]

Luckily for the future, de Vaucanson didn't stay a monk for long. At age eighteen, the order had set him up in his own shop in Lyon with a sizable contract from a local nobleman to build machines. Learning that he would soon be visited for dinner by one of the heads of the order, de Vaucanson decided to make the meal itself a demonstration of what he could accomplish.

So, de Vaucanson built a collection of humanlike automatons (or, more properly, androids) to serve the dinner and later clear the table. The visiting dignitary was politely impressed during the meal . . . but soon afterward declared de Vaucanson's activities to be "profane" and ordered the workshop destroyed. It wouldn't be the last time that de Vaucanson would be attacked for violating the status quo.

De Vaucanson was clever enough to see where this was going. So he went home to Grenoble, begged the bishop for a release from his vows because of an "unmentionable illness," and once he got approval, ran for Paris. There, like a good entrepreneur, he built prototypes, found some angel investors, and located subcontractors to do his manufacturing. Soon, also like a good entrepreneur, he found a venture capitalist to underwrite taking his devices on tour, and started making some real money.[13] At this point de Vaucanson might have enjoyed a very profitable career building watches and novelty automatons.

But then everything changed. He fell deathly ill and was bedridden for four months. In his delirium, a vision came to him of a full-sized human automaton, in the shape of a currently famous marble statue by sculptor Antoine Coysevox, which could play the flute. As legend has it, the moment de Vaucanson recovered, he quickly sketched out all of the components of what would be called "The Flute Player" from the complete designs in his head. Teams of craftsmen scattered to start building, and when the finished parts were at last assembled the Flute Player worked perfectly.

With one exception: The flute being an incredibly delicate instrument to play, de Vaucanson found that he had to cover the automaton's articulated wooden fingers with real skin—leading a later observer to comment, only half jokingly: "What a shame the mechanician stopped so soon, when he could have gone ahead and given his machine a soul."[14]

DUCK TALE

The Flute Player, made of wood but painted white to resemble marble, debuted in a private gallery on February 11, 1738. The automaton stood about five-foot-six, its height nearly doubled by a large pedestal to make it look even grander. Only about a dozen people were allowed into the gallery at a time, and de Vaucanson charged each 3 livres—nearly a hundred dollars in today's money. No one seems to have complained about being overcharged; on the contrary, most were utterly dazzled by the the Flute Player, whose repertoire included twelve different melodies. It also caught the eye of the aristocracy, in particular, the court of Louis XV.

Such was the sensation of the Flute Player that de Vaucanson probably could have retired a very young and very wealthy man. But when the crowds began to thin for the Flute Player, the driven inventor was already prepared with a follow-up: an equally large, tambourine-playing automaton in the form of a shepherd. It proved a success as well. More important, it paved the way for de Vaucanson's final masterpiece: the Digesting Duck.

Not only did the Digesting Duck represent a huge leap in the history of machinery (its frauds aside) but, as already noted, it captured the fancy of the leading lights of France and much of Europe. Frederick II of Prussia invited de Vaucanson to join his court—an offer the patriotic Frenchman refused. As Voltaire famously joked, "What better image of the glory of France than a shitting duck?"[15]

Having tickled the fancy of the monarchy, de Vaucanson was now rewarded in their typical way. In 1741, Cardinal Fleury, chief minister of Louis XV, appointed de Vaucanson, the glove maker's son, to the post of inspector of the manufacture of silk in France. Presumably, even though it was awarded by the famously pragmatic cardinal, this was to be a largely ceremonial sinecure. But the thirty-two-year-old de Vaucanson took it as nothing of the sort. In short order, he sold off his great automatons and set to work revitalizing France's troubled silk industry. He was done forever with automatons; now he would automate the world.

SILKY SOLUTION

In the mid-eighteenth century, French silk manufacturing, once one of the country's most powerful industries, had fallen on hard times. Like successful industries even today, it had been a victim of its own success and had grown wary of change, resistant to upgrading its expensive capital equipment, and content to pull high profits out of this cash moth. As a result, it was ceding market share, decade by decade, to newer, hungrier, and less risk-averse competitors in England and Scotland.

De Vaucanson seemed to instantly understand that what the French silk industry needed to get back into the game was to technologically leapfrog its rivals. Toward that end, in the decades that followed he would himself invent a number of important machine tools—most famously the metal slide-rest lathe, an invention deemed so important at the time that Diderot and his team put it into the *Encyclopédie*.

But de Vaucanson's greatest invention—both for French silk history and human prosperity—was his first: the automatic loom.

As far back as 1725, a couple of inventive French weavers, Basile

Bouchon and Jean Falcon, had independently tried to make the process of weaving more systematic. Bouchon, in particular, had come up with a device that used a perforated paper tape to control the raising and lowering of the needles that regulated the height of the warp threads as they were woven across the weft surface of the fabric—a process that had traditionally been done using a series of cords. The invention had only proven marginally successful: Reliability problems aside, it also required yet another worker on hand to feed the tape.

Twenty years later, looking for a breakthrough technology, de Vaucanson seized on the paper-tape idea—and then applied to it the genius of the world's greatest automaton maker.

Most inventors are celebrated for making a single design breakthrough. But with his loom alone, de Vaucanson made *three*. First, he not only recognized the potential power of Bouchon's now nearly forgotten invention but also saw its key weakness. Second, seeing there was an analogy of the paper tape to the pinned barrels of his own inventions, de Vaucanson essentially took a barrel, sliced it up one side, and flatted it out into a durable wooden "card" about the size of a modern automobile license plate. Then, taking Bouchon's cue, he replaced the pins with drilled holes. Third, and most important, he strung the cards together side by side and—like the reduction gear on his automatons—created a controlled feed for them into the loom, powered by the loom itself and synchronized to the weaving.

In operation, with each line of the weave, the warp needles would push up against these cards, penetrate through wherever there was a drilled hole, and lift the warp thread to the surface. The result, we recognize today, was a binary event: on or off, closed or open, go or stop. To change the pattern of the finished fabric, you merely had to swap out the cards.

What de Vaucanson had invented was the world's first fully automated, user-programmable production machine.

POINT OF INFLECTION

It can be said of few men or women who have ever lived that they materially influenced a historic epoch. In de Vaucanson's case, he influenced *two*. His loom stands, along with the steam engine and the Bessemer steel furnace, as one of the defining inventions that made the Industrial Revolution possible. But it is recognized as a key source of the subsequent Information Revolution as well. That's because once you replaced the needles going through those holes with electricity, you could build computers and write complex computer instruction code: software. Use that same on-off screening with light and the light-sensitive emulsion that would be invented in another century, and you had photography, photolithography . . . and ultimately the semiconductor chips that would fill those computers.

What's more, though de Vaucanson didn't recognize it at the time, his automatic loom had yet one more discovery hidden inside of it. If he had ever attempted to reverse the loom's operation—that is, to use the needles to extract the location of each line of thread and then feed that backward into a device that *drilled* the holes in the cards—de Vaucanson would not only have invented instructional memory but memory "recording" as well.

Unfortunately for M. de Vaucanson, whereas his clever automatons had provided a thrilling after-dinner entertainment for rich royals, his far-more-important automatic loom only managed to provoke anger among the working classes, who saw the new invention as a threat to their livelihoods. Presaging the Luddites, French silk workers not only attacked the looms, but de Vaucanson himself, showering the inventor with stones when he visited their factories.

It would be fifty-five years before another intrepid inventor, Joseph Marie Jacquard, would revisit de Vaucanson's design and perfect the loom that forever bears Jacquard's name. And, with no little irony, the Jacquard loom would find itself most valued in England, where recent inventions in thread-making, such as the Spinning Jenny, had created the need for a fast, configurable, and automatic loom. In the end, de Vaucanson's loom not only failed to restore the French silk industry, it gave its British competitors the missing piece it needed to capture the entire industrial world.

The Industrial Revolution, which the Jacquard loom helped to power, is—as every schoolchild is taught—one of the two or three most important points of inflection in human history, ranking right up there with the birth of agriculture. But living as we do on the far side of that revolution, and knowing only a world irrevocably changed by it, it can be difficult to completely appreciate the full measure of this transformation.

But if you look at historic graphs of human demographics and behavior, nearly every one is essentially a straight line (with a few brief anomalies) for five thousand years beginning with that birth of agriculture. Life expectancy didn't much change, infant mortality stayed high, and educational levels remained low. Per capita income? About the same. Average speed of movement? Walking, with bursts up to twenty miles per hour with horses and sails. Regularly recurring plagues? Check. Famines? Check.

In other words, even if you include the genius of Periclean Greece and the Italian Renaissance, the might of Imperial Rome and the Persian Empire, or the endurance of Dynastic Egypt or Imperial China, the lot of mankind changed little over the millennia. The notion of progress, which defines every minute of every day of modern life, was all but nonexistent before the Industrial Revolution and usually reserved only for making one's place in heaven.

And then suddenly, shockingly, around 1800 all of those flat lines curve upward. Humanity starts living longer, is healthier and bigger. More children survive—and go to school. Plagues become less frequent and (with one big exception) less murderous. Per-capita income skyrockets as more and more people leave the once-dominant profession (farming) and flock to the rapidly growing cities to take better-paying and more reliable factory jobs, the requirements of those jobs in turn driving literacy levels to nearly 100 percent. This is turn precipitates the greatest jump in productivity in human history, increasing both personal wealth and ownership.

With people healthier, living longer, and having more surviving children, population shoots upward as well. And these new uprooted populations also begin moving about more quickly: The locomotive presented the first real improvement in human speed since the domestication of the horse 20,000 years before.

Yet for all of this, the single greatest change wrought by the Industrial Revolution may have been our relationship with time. The timing and sequence that defined de Vaucanson's automaton was now mapped onto all of civilization, and with each decade time marched faster and the sequences grew ever more sophisticated. The English system of factories, steel, and improved transportation quickly made Great Britain the most powerful and wealthy and country in the world—and set off two centuries of industrial espionage as other countries raced, by any means necessary, to duplicate its success.

None was better at this than the United States—and the American scheme of systematization of processes and interchangeable components would in turn set off what would be called the Second Industrial Revolution of large-scale mass production that would bring the fruits of manufacturing to nearly every person on the planet. Automatons had now become *automation*, and human beings were not only the observers and consumers of this newly automated world, but they were now also components *within* the automaton itself. It's not a coincidence that early in this process Mary Shelley created that most influential of all modern myths, *Frankenstein*—the horrifying tale of a man beset by a monster of his own making.

Indeed, the role of people in this newly industrialized society became problematic. The task of one group—investors, bankers, managers, scientists, experts, and eventually marketers—became that of making the machines more powerful, productive, and cost-efficient. For the other group—workers—the job was to become a component inside the machine, moving the process along at pivot points where automation couldn't be trusted to accomplish the work itself.

As we all know, for the latter group, all of the advances in wealth and quality of life wrought by industrialization didn't make up for the dreariness and anonymity of assembly-line work. And this frustration reached a breaking point when, in order to keep up with the ever-faster machines, timing and sequence improvement evolved into a science: Frederick Taylor's "scientific management," with its time-motion studies and obsession with efficiency. The resulting explosion in productivity led to modern professional management, with its focus on empirical decision-making and a shared collective memory of "best practices."

But before that, the revolt against automation gave us the labor movement, with its reassertion of the rights, the dignity, and most of all, the *humanity* of those workers trapped inside the automaton. The ghosts had at last fought back. And thus, it shouldn't be surprising that perhaps the most resonant image of this revolt was Chaplin's Little Tramp as a factory worker in *Modern Times,* being processed through a series of giant gears—the simplest form of pinned barrel—like, well . . . shit through a duck.

Somewhere, Voltaire was laughing.

Jacques de Vaucanson spent the rest of his life inventing and earning awards, including membership in the French Academy of Science. But history soon forgot him and his remarkable contribution. There is an engraving of de Vaucanson, made just before his death, showing a bewigged old man with the face of a wary bloodhound.

De Vaucanson died in 1792. He didn't live long enough to see either the revolt of his factory-worker employees over his former royal audiences (and the destruction of many of his automatons) or the other great social revolution his inventions helped to create. And if, in the end, he hadn't managed to bring his machines to life, de Vaucanson had done the next best thing: He had given them independent lives of their own.

8
Tick, Talk
Memory as Recording

In the century between 1870 and 1970, the world of memory was dominated by a new kind of figure: the entrepreneur/inventor.

Memory work had heretofore been the province of scholars, ecclesiastics, and scientists. But as the focus of memory shifted from the human brain to mechanical systems, and as command economies transformed into free-market capitalism, the nature of the players changed as well. The goal now was not just to develop new memory technologies but to turn them into successful and profitable enterprises. If lost in this transformation was research done for its own sake, what was gained was rapid innovation, continuous improvement forced by competition, and, through interaction with users, practical results.

The century after the American Civil War offered a brief window of opportunity for these singular individuals. Entire markets lay open for prospecting, inventions could still be made by single individuals, and manufacturing was cheap. Moreover, the public, increasingly trained to become enthusiastic early adopters of new inventions, not only provided a perpetually hungry market but was happy to lionize these inventors as heroes of the age.

And they *were* heroes. But they were also geniuses, eccentrics, megalomaniacs, and, more often than not, doomed by the same traits that had made them successful in the first place (stubbornness, single-mindedness, and contempt for the work of others) to ultimately destroy their creations.

But before they did, they changed the nature of memory, human and artificial, forever.

GO-GETTER

John Shaw Billings was a classic "Can do!" character of post–Civil War America, one of those men who turned the United States from a war-battered frontier nation to a global empire in the span of a single generation.

With his walrus mustache and baggy eyes, Billings looks slightly embarrassed by all the attention in his portraits and photographs, perhaps in a little over his head. But he was anything but: Billings was, in fact, brilliant, pragmatic, a veteran leader and administrator, and utterly unafraid to take on the biggest and most thankless tasks.

Like many of his successful contemporaries, Billings had come from humble beginnings in small-town antebellum America and had made his success through brains, pluck, and hard work. In Billings's case, he was born in 1838 in Allensville, Switzerland County, Ohio. He graduated from Miami (Ohio) University in 1857 and completed his medical degree three years later at what is now the University of Cincinnati College of Medicine. Thus, he was just a few months into his practice when Fort Sumter was attacked and the nation, North and South, mobilized for war.

Following enlistment, Billings found himself in Washington, D.C. The Union Army was facing huge problems keeping up with sanitation (disease would kill more soldiers than battle would in the Civil War) and with equipping and staffing both field surgeries and military hospitals. It was one of those moments when merit is actually recognized and rewarded: At age twenty-four, Billings was named medical inspector of the 200,000-man Army of the Potomac. By all accounts, he served brilliantly.

Following the war, Billings was appointed librarian of the Surgeon General's Office—a job with more title than substance. Billings changed that: The collection he compiled became the heart of the National Library of Medicine.

Though he held the title of librarian for thirty years, it didn't keep Billings from taking on other "impossible" tasks whenever they were offered. For example, he was also professor of hygiene at the University of Pennsylvania . . . and while there he played a key role in designing all of the original buildings for Johns Hopkins University (its well-known dome is named for him). Later in his career, Billings was invited to New York City to pull together the local borough libraries into the New York Public Library system, still one of the world's largest.

It says something about Billings's reputation and his comfort with power that it was he who first approached, and then convinced, tycoon Andrew Carnegie to build sixty-four new branch libraries in New York City, and then—in probably the greatest individual act of philanthropy in modern times—to build three thousand public libraries in cities and towns across North America and the British Empire. Most of these "Carnegie" library buildings still stand, like secular temples, in small-town America, Canada, the United Kingdom, Australia, New Zealand, Fiji, and the West Indies—and they played a major role in raising literacy rates in many of those nations.

But it was in the mid-1870s when Billings received, and accepted, what he believed would be the biggest challenge of his career: directing the 1880 U.S. Census.

THAT ALL BE COUNTED

The census was hardly a new idea. Counting populations had already been going on for five thousand years starting with Egypt, the first believed to have taken place in 3340 B.C. during the early Pharaonic period. Within a millennium after that, the Chinese were conducting a regular census, and some of their records still survive to this day. The Jews in Israel also conducted a regular census for tax purposes (as commanded by God in Exodus), a process so embedded in the culture that it even gave name to the Old Testament book of Numbers.

Rome reportedly began taking a regular census as early as its pre-Republican kings, in particular Servius Tullius in the sixth century B.C. By the time of the empire, census-taking had become a massive undertaking stretching across Europe and the Middle East—an event memorialized in the opening of the Gospel according to Luke, where Joseph of Nazareth and his pregnant wife, Mary, have to travel to Bethlehem to be recorded and taxed. Despite the immense challenge, Rome considered its census to be so important that the empire ran it as a continuous program for centuries, adding the results every five years. "Census" is, in fact, a Latin word for "assessment," the measure of all male citizens.

Census-taking was pretty much abandoned in Europe during the chaos of the Dark Ages, but it continued in the Middle East under the caliphate, beginning under the caliph Umar in the early seventh century, soon after his conversion to Islam. But almost as soon as order was restored to Western Europe, population counting began again in earnest.

The most famous census of the Middle Ages was the *Domesday Book*, ordered by William the Conqueror in 1086 to survey his newly conquered British Isles and to determine what taxes and tributes had been owed to his predecessor, Edward the Confessor, and that now belonged to him.

According to the *Anglo-Saxon Chronicles*, the idea came to William the previous year:

> *While spending the Christmas of 1085 in Gloucester, William had deep speech with his counsellors and sent men all over England to each shire to find out what or how much each landholder had in land and livestock, and what it was worth.*[1]

It is often assumed that the *Domesday Book* is simply the record of that census. But it is not that simple. In this world before printing, of vellum and with shortages of scribes, the census *was* the book—and the book was the *law*. Hence the name. To the Anglo-Saxon citizens of England, under the heel of their new Norman overlords, this thick volume really was the "Doomsday" book—the book of Last Judgment. Against its records, there was no appeal.

By the sixteenth century, as part of the zeal to catalog and enumerate that characterized the Renaissance, censuses were taking place in almost every kingdom and principality in Europe. They were also being held in the Middle East, in China, and even in South America in the Incan Empire. In 1183 the Crusaders in Jerusalem undertook a census to determine if they could raise a sufficient army to defend against attack by Saladin. In 1577, King Phillip II of Spain ordered a census of the entire Spanish empire in the Western Hemisphere to determine Spain's total wealth and population of captured peoples (Spain itself had held censuses since the thirteenth century). In 1666, France did the same with its colony in Canada; and the British raj did so at last in India in 1872.

As these examples suggest, the appeal of census-taking to rulers and legislatures took many forms. For the pharaohs, it was a way to measure their own power in people, structures, and slaves. For the Caesars and the emperors of China, it was all of that, plus a means to determine the potential number of men-at-arms to fill armies. And, of course, by the Middle Ages, it was the primary tool to squeeze taxes out of the citizenry.

But the censuses were also something else that was largely lost on the people of their time but is now their primary value: They were the ultimate (and often only) memory records of *everyday people*.

Of the ancient world, and even as late as the Renaissance, we know almost nothing about individual shopkeepers, laborers, and soldiers—much less women and slaves—unless they did something extraordinary, which by definition made them more than ordinary people. There was no photography, and in the lower classes there was general illiteracy, so it was almost impossible for any person outside of the ruling classes to leave even the slightest record of their existence for future generations.

As a result, we have almost no records of what it was like to be a laborer on the pyramids or a low-ranking Roman legionnaire, a Chinese farmer during the Tan dynasty, or a Greek hoplite. These memories will never be recovered because they were never recorded. The memories of a billion people over the course of a thousand generations are lost to us . . . forever.

Censuses changed that. Not right away, of course. Rulers cared little about the biographical details of their plebian subjects, and not at all about the proletarians and slaves. But that all changed when the middle classes began to emerge across the Continent and governments suddenly became interested in individuals, their extended families, and most of all, the ownership of their assets.

NEW WORLD, NEW MEASURE

For a newly emerged democratic republic such as the United States—and a nation that wouldn't have a federal income tax for its first century—a census served a wholly new set of purposes: to establish voter registrations and apportion districts; to create rolls of adult males for potential military drafts; to locate the ever-growing numbers of new immigrants; and to count the number of slaves and indentured servants (though the former would only be officially recorded as three-fifths their total number). But perhaps the most important task in a nation that was rapidly expanding its borders westward was to locate and count the hundreds of thousands of citizens spread across the wide frontier—some of them so isolated that they continued to vote for Andrew Jackson for president long after he had served his terms and died.

The different American colonies had held their own censuses since the seventeenth century, the first taking place in Virginia at the beginning of that century. The first official U.S. Census took place starting on August 2, 1790, when the nation was officially just one year old. To underscore how seriously the federal government considered the task—and to give a classic example of President George Washington's decisiveness in establishing executive precedents—federal marshals, under the command of Secretary of State Thomas Jefferson, were sent out to knock on every door, take down the name of every head of household, and then count the number of other individuals in the residence.

Interestingly, while slaves were counted in this first census (and, as noted, only as fractional citizens for apportionment), Native Americans were not—both because they weren't considered "real" citizens and because of the potentially mortal risk to the census takers.

The final tally of this first U.S. Census was 3.9 million Americans—a reminder of just how tiny the nation was that had just fought its way to independence from history's largest empire.

The second U.S. Census took place in 1800, thus setting a pattern of ten years between counts—a very brief interval, given the growing effort required, but almost too long for a nation growing as rapidly as the United States. Meanwhile, each census added a few more questions as the young government struggled to understand how its citizens lived, where they worked, and the forces that underpinned the national economy.

Thus, as early as the third census in 1810, questions were already being asked about manufactured goods. The 1840 census added questions about the nation's fisheries, perhaps reflecting maritime disputes of the era.

But it was with the 1850 census—perhaps reflecting the growing schisms of a nation sliding into Civil War—that the U.S. government went all the way in. Now it asked questions about taxes paid, criminal records, religious affiliation, and even whether the subject was destitute. Moreover, for the first time (and to the eternal gratitude of future genealogists) the census takers also asked for the complete names and ages of *all* household members. And in another unprecedented turn, even family slaves were fully named.

Now, for the first time in history, the average person had a name, a family, a way of life . . . and a memory that would live long beyond them, to be read by their descendants centuries into the future. Even more remarkably, slaves, the ultimate forgotten figures of history, at last entered into human memory—a harbinger of the Emancipation Proclamation still more than a dozen years away.

ADDING UP A NATION DIVIDED

If the 1850 census had predicted a brighter future, its 1860 successor took that hope away. It was, as Adam Goodheart dubbed it 150 years later in *The New York Times,* "The Census of Doom."[2]

By this point, the act of taking a U.S. Census, compiling the data, and then publishing the information took almost five years. Needless to say, with the nation coming apart at the seams—and just months before the outbreak of conflict—there was no time nor funds for a fully elaborated report like that of a decade before. By the time the results were finally tabulated, fighting had begun, and so census superintendent Joseph Kennedy, working with a staff of only 184 clerks to count the millions of records, ultimately delivered only an abbreviated report that lacked the expected cartographical representations of where America's populations were located.

Still, enough data was processed and made public between the June 1860 start date of the census and the November 1861 presidential election to have a profound effect on the lead-up to war—and thus the future of the United States. Wrote Goodheart:

> *Preliminary figures that began appearing in the press as early as September 1860 confirmed what many Americans already suspected: immigration and westward expansion were shifting the country's balance of population and power. Since the last count, in 1850, the North's*

> *population had increased an astonishing 41 percent, while the South's had grown only 27 percent. (Between 2000 and 2010, by comparison, the entire nation's population grew just 9.7 percent.) Tellingly, the statistical center of national population had shifted for the first time not only west of the original 13 states, but also from slave territory into free: from Virginia to Ohio.*[3]

These numbers shocked the Southern slave-holding states. The survival of slavery had long depended on a voting balance between North and South, by which the states below the Mason-Dixon line could hold off the abolitionists above it. Now it was obvious that this fragile balance had been shattered—and the South would never again have control over its own destiny. Not only would these booming populations all but decide future presidents, but reapportionment would soon be adding new congressional districts to the North, while taking them away from the South.

And, at least for Southern whites, it only got worse. Not only were the populations of the Northern states booming but so were those in the frontier regions of the old Northwest Territory, the Great Plains, the Southwest, and the Far West—territories that, thanks to their "free-soil" pioneers, were all likely to vote anti-slave with statehood.

Of all of these territories, Wisconsin and Minnesota represented the worst Southern nightmare:

> *In 1836, [Wisconsin] had claimed fewer than 12,000 inhabitants. Now, in 1860, it boasted 778,000—an increase of almost 6,400 percent in less than a quarter of a century . . . Nor was it even the most remarkable case. Neighboring Minnesota's population had risen from 6,000 to 172,000 in the past decade alone.*[4]

Making matters worse, in Southern eyes, the inhabitants of these two states were largely immigrants, Northern Europeans who had no history of slavery and who had a particular hatred of the Peculiar Institution. Even the South's long-standing argument that slavery was too important an economic institution to dismantle was now beginning to lose its credibility: The census data also showed (and *The Philadelphia Inquirer* was happy to point out) that slaves now, at just 12.5 percent, represented the lowest percentage of the U.S. population since measurement began.[5]

None of this was lost on the South. The writing was on the wall: Slavery, the largest single industry in the United States, was doomed. And since slavery was the underpinning of the entire Southern agrarian culture, so, too, was it the heart of the Southern economy. At the same time, white Southerners, who had led the nation through the Revolution and the early years of the Republic, were now destined to become a minority, slowly stripped of their government representatives, their role in American life supplanted by immigrants in the North.

If the 1860 Census didn't cause the Civil War—those roots reached back to the treatment of slaves in the Constitution—its data most certainly hastened its arrival. Southerners, seeing these results, now looked on with greater dread than ever at the coming presidential election and its Republican, Northern, and antislavery candidate, Abraham Lincoln. It is a measure of just how frightened the South was of these shifting demographic winds that Lincoln won the presidency but failed to win a single electoral vote from the Southern states. The stage was set; there was no going back to the status quo. Using the same census data, Southerners convinced themselves that because they were now a larger population than the American colonies at the time of the Revolution, they, too, could shrug off a mighty continental empire.[6]

Once the Civil War did break out, the Census Department quickly became an arm of the War Office—in particular, Kennedy and his team, now armed with the best population data in the country, began creating a new set of population maps for field commanders and military governors. These maps included information not just on white populations but slaves as well (and even Indians, or at least those 40,000 Native Americans who had "renounced tribal rules"), along with local agricultural products and, crucially, train routes and schedules.

The total population of the United States according to the 1860 Census was 31,443,321. Within five years, more than 620,000 of those citizens would be killed.

The next U.S. Census, in 1870, landed smack in the middle of Reconstruction in the South, the Gilded Age in the North, and massive pioneer emigration to the Great Plains and beyond. America's official population was just short of 40 million.

THE LONG COUNT

In 1880 it was John Shaw Billings's turn to count all Americans. By all accounts, he did the job brilliantly, despite dealing with a much larger (50 million, having jumped by 30 percent in a single decade) and more geographically diverse population, and a set of census questions of unprecedented size. Indeed, the 1880 census had five separate schedules (sections), with only the first resembling the traditional headcount that could be filled out by citizens. The other four sections—mortality, agriculture, manufacturing, and social statistics—asked dozens of questions about marital status, birthplace and cause of death of parents, crops and their rotation, fertilizer purchases, number of seasonal hired hands, average daily wages, capital equipment, debt, education, and a host of other matters . . . all of which had to be asked by census field "enumerators." Special customs agents also went into the field to gather statistical data on all of the nation's major manufacturing industries.

It was a gigantic effort—and in the end, an extraordinary achievement. For the very first time, a great nation fully understood its own nature . . . and prepared the most complete memory of itself for future generations achieved to date. But it hadn't come easy, despite the fact that Billings had broken with precedent and for the first time hired women to join his army of enumerators. In the end, it took *seven* years to complete the 1880 U.S. Census.

The 1890 census loomed just ahead. And John Shaw Billings knew he had a serious problem on his hands. By then there might be more than 70 million Americans, filling even more of the hard-to-reach corners of the nation. And the planned census schedules were even more complicated in their questions. The 1880 census had taken seven years; Billings had only to extrapolate out to realize to his horror that it would take an estimated *thirteen years* to complete the 1890 census as it was currently configured. Until then, the entire U.S. government—and likely the economy itself—would be working blind on wildly obsolete data. The results could be catastrophic.

There was no way that one census could be allowed to overlap the next. After such an illustrious career built on an unmatched reputation for competence, Billings realized he was facing career suicide and public humiliation.

But John Shaw Billings wasn't the type of man to shrink from challenges. As he'd shown when he hired women, he saw only the task to be done, and was prepared to crash through any technical or cultural barriers in his path to getting there.

He had been watching one of his employees—an eccentric, young former mining engineer named Herman Hollerith—who seemed to have a gift for solving mechanical problems. And it was over a chicken-salad dinner in 1879, not long after Hollerith joined the Census Bureau, that Billings first raised the challenge of speeding up the 1880 census and the looming specter of the 1890 census beyond. Hollerith left that dinner driven to find a solution.

He didn't make it in time for the 1880 census, but Herman Hollerith was not a man to give up on a challenge.[7] Born in Buffalo, New York, in 1860, he had moved at an early age with his German immigrant parents to New York City. There, young Herman proved to be a difficult student—brilliant in the subjects he cared about but indifferent, even truculent, about those he did not. Legend has it that he once even jumped out of a schoolroom window to escape a spelling test.

Eventually, his despairing parents pulled Herman out of school and put the young man in the hands of a private tutor. Hollerith bloomed under the customized curriculum and was able to enter Columbia College at just sixteen. Three years later he graduated as a certified mining engineer. Ironically, given his future, his two worst subjects were machinery and bookkeeping. Taking a job at the Census Bureau, he moved to Washington, D.C., and within a few weeks he had his fateful dinner with John Shaw Billings.

Like many geniuses, Herman Hollerith's talent lay not in creating something wholly new, but rather in visualizing something that already existed in a wholly different context. Riding the train back and forth to New York City, Hollerith couldn't help noticing the railroad conductor passing from seat to seat, using a handheld punch to put holes in passengers' train tickets in specific locations to designate gender, approximate age, and other characteristics to prevent fraud. These punched tickets were crude descendants of the punched cards used in jacquard looms, and thus owed their existence to de Vaucanson's Digesting Duck.

In the intervening 150 years, the notion of using pieces of memory for the purposes of controlling systems had spread from industry to daily life—and from automatons and looms to train tickets, employee "punch" time clocks, player pianos, and music boxes. In operation, most of these were very simple compared to the sophisticated and "programmable" looms now in use around the world. For example, the "machine" in the punch-ticket application was the conductor himself, whose job it was to compare the traits marked on the ticket with the passenger seated in front of him. But nevertheless, together these various applications had begun to create a growing regime of machinery control through data entry.

Millions of people now experienced one or more of these control devices almost on a daily basis. But only Herman Hollerith had the combination of genius and the knowledge of a pressing need to see how these systems might be *reversed*—that is, instead of encoding commands, they could compile results.

Throughout the 1880s, Hollerith, with Billings's support, began to bring together various existing technologies in a new design of his own, which he called a *tabulator.* Billings, it should be said, while a staunch supporter of his subordinate's work, was also far too pragmatic to pin all of his hopes on the odd twenty-five-year-old. So, even as Hollerith labored away, Billings also put out a public call for entries in a competition for a new census-compiling machine, with a deadline of 1887.

In the end, there were three entries. In addition to the design of Hollerith, who had finished building his tabulator just in time, there were also working models from two other inventors, Charles F. Pidgin and William C. Hunt. Each machine in turn was run through a representative set of data from the 1880 census, a task that mostly involved converting handwritten census data into machine coding, then slicing and dicing the entered data for various categories of results.

When the results were scrutinized, it was discovered that the Hollerith tabulator had not just defeated its two competitors but blown them away. Hollerith's machine proved to be twice as fast as the Pidgin machine, and three times as fast as the Hunt. This also meant that the tabulator was probably at least ten times more efficient than traditional handwork.

The tabulator design had triumphed because Herman Hollerith had not just focused on one big breakthrough—though he had one in the design and use of inexpensive punched paper "cards"—but also on bringing together diverse technologies to take full advantage of that core innovation. Thus, the tabulator had a means of mechanically loading new cards into the machine, where they could be punched using the new typewriter technology. Then, after the new holes in the 12-row-by-24-column face were electrically "read" by spring-loaded copper wires dragged across their surface to make contact with a metal conductor placed below, they could be mechanically sorted by their common data into stacks and dropped into a series of boxes while an overhead bell rang to signal success. This process could then be reversed, through a counter, to add up the results, which appeared on dials, one for each decimal place, on the display. And even though many of these steps were still performed manually, the process was still faster than any data storage—that is, artificial memory—ever before attempted.

COUNTING COUPS

John Shaw Billings, fearless as ever, immediately committed the entire U.S. Census Bureau to the Hollerith tabulator and, beginning in June 1890, these unproven machines held the collective memory of the United States in its gears, keys, and cables.

After just *six weeks* of processing—instead of the previous two years—the Census Department announced its preliminary results: the United States now had a population of 62,947,714.

The news shocked many Americans, especially those in positions of power, for several reasons. First, the results, thanks to the Hollerith tabulator, had been announced so quickly—it seemed impossibly quickly—that there was bound to be questions about the quality of the work. Had the Census Bureau taken shortcuts? Grown sloppy? Or even *made the data up*? A headline in the *New York Herald* read:

SLIPSHOD WORK HAS SPOILED THE CENSUS

MISMANAGEMENT THE RULE

Speed Everything, Accuracy Nothing![8]

Only adding to these suspicions was that the population figure seemed unexpectedly low. The United States had been growing, the past censuses said, at about 25 percent per decade for most of the nineteenth century. This suggested a likely 1890 population figure of 65 million and, given the huge influx of immigrants in recent years, some analysts had predicted the number might even reach 75 million. This apparent shortfall, combined with the unbelievable speed with which the figure had been reached, led the skeptical, especially those with a political stake in the results, to suggest more nefarious motives.

But a confident John Shaw Billings ignored the naysayers and plunged on. The 1890 census asked questions in twenty-five categories, including nearly all of those from the 1880 census, adding queries regarding citizenship and naturalization, chronic illnesses, permanent physical defects, months unemployed in the last year, farm ownership, Civil War veteran's status, and finely detailed questions about race (specifying categories like mulatto, quadroon, octoroon, and so on).

The federal government fully expected that the processing of all of this data would have taken, if not as long as Billings's worst nightmare of thirteen years, then at least almost the entire decade. So one can imagine the shock when the Census Bureau released all of these results, along with maps and other supporting documents, in just *one year.* This time, the complaints were muted: The details of the census report were now just too complete to deny. Billings and his team seemed to have performed a miracle.

One effect of this remarkable achievement was to make both the Hollerith tabulator and Herman Hollerith himself the subjects of considerable demand around the world. Hollerith, like many corporate and government "intrapreneurs" to follow in the next century, quit the Census Bureau and started his own company to build new, upgraded tabulators. He received his first international order from Russia for its 1897 census. Orders from other European nations quickly followed. The U.S. Census Bureau also ordered Hollerith's new models—the integrating tabulator, which could not just count but also add numbers represented by punched holes on the cards; and, in time for the 1900 U.S. Census, an automatic-feed tabulator to further automate the process.

Now forty years old, Herman Hollerith was becoming a very wealthy man. But as he began to indulge a growing taste for cigars and fine wine, he lost none of his eccentricity. When asked, in the early days of his company, why he had not yet applied his tabulators to the lucrative field of railroad accounting, he replied (as he later told others): "One good reason, and that was that I did not know the first damned thing about railroad accounts."[9]

Nevertheless, within months, Hollerith was delivering his machines to the company's first private customer: the New York Central Railroad. And in the years that followed, he added new customers at other railroads, public utilities, and even one department store. But without a John Shaw Billings to manage and direct him, Herman Hollerith proved to be a much better inventor than businessman. Billings had left the Census Bureau and taken on a new role as the federal government's investigator/liaison to the women's temperance movement. Dying in 1913, he didn't live to see the bittersweet results of his last successful project: Prohibition.

Hollerith had one last triumph in 1906 with a new Type 1 Tabulator. It featured an added wiring panel that enabled the tabulator's operation to be modified for different tasks, making it one of the key milestones along the path to the invention of the computer. But by then it was too late. The year before, a frustrated U.S. Census Bureau had given Hollerith an ultimatum to upgrade the tabulator *and* reduce his fees. Being Herman Hollerith, he refused—then went ahead and built the Type 1. But the Bureau had had enough; it dropped the Hollerith contract and started building tabulators of its own. The subsequent lawsuit lasted seven years and ended with a victory for the government.

By then it didn't really matter: Hollerith had sold his company to a newly formed conglomerate, Computing Tabulating Recording Corporation (CTR), with which he retained a major stock position and the title of chief consulting engineer. It was a role that suited Hollerith, who was weary of business and increasingly drawn to a simpler life of farming and cattle breeding.

In 1914, entering into Herman Hollerith's life was a man who was as quintessentially an American businessman of the twentieth century as John Shaw Billings had been a government bureaucrat of the nineteenth. Thomas Watson Sr. was a gifted salesman on his way to becoming industrial titan, and had Hollerith followed his advice the way he had that of his old boss, his profile in history would have been far higher.

Watson knew what his customers wanted: permanent memory output. In this case, in the form of a paper tape printer. It would have been easy for Hollerith to modify his existing Type 1, but instead, he refused to have anything to do with the work—not least because he despised Watson and all that he represented in the growing role of corporate sales and marketing. Instead, Hollerith retired to his farm and the simple, good life.

Watson, meanwhile, got his paper tape machine—and soon added printers and removable plug boards. And in 1924, he changed the name of CTR to International Business Machines . . . IBM.

THE WIZARD

While Herman Hollerith was struggling to build his first tabulator, the world's greatest inventor was struggling to consolidate his success of the previous two decades.

Thomas Alva Edison, whose inventions had transformed the world, had hit a rough patch in his life. Mary Edison, whom he'd married in 1871 when he was twenty-four and she a sixteen-year-old employee in his shop, had died of a brain tumor in 1884, leaving him with three children. And Edison himself had spent much of the 1880s battling one competitor after another in courtrooms and in the marketplace.

He had been sued by Emile Berliner over the patent to the carbon microphone—the heart of the telephone for the next century—and eventually won. He had battled George Westinghouse over control of all electrical power distribution in the United States, Edison stumping for his direct-current (DC) design against Westinghouse's promotion of Nikola Tesla's alternating-current (AC) design. (Edison eventually hired Tesla and then managed to turn the mysterious scientist into a lifelong enemy.) That fight, with so much at stake, had quickly turned ugly, with Edison employees publicly electrocuting animals—most notoriously a circus elephant—and filming the results to prove the dangers of AC.

Meanwhile, in his rush to develop the fluoroscope as an efficient diagnostic tool for the use of the newly discovered X-rays, Edison had accidentally given one of his favorite employees (who had been a ready volunteer) a mortal dose of radiation.

And most depressing of all, Edison had been sued over his most famous invention, the incandescent lightbulb. After his patents in telegraphy, the lightbulb had been the major source of Edison's wealth—and even those revenues had been under assault since 1882, when George Westinghouse had bought the patent to the rival "induction" light, slashed licensing fees, and in the process forced Edison to cut his own patent fees. Then, a year later, the U.S. Patent Office stunned the world by awarding the patent on incandescent light to William Sawyer, rendering Edison's patent invalid. It took six years of litigation before a judge officially recognized the superiority of Edison's claim.

Edison had come a long way from the single-minded young inventor of the 1870s who had forever transformed the world with a dazzling run of new creations, including the automatic ticker tape machine, the mechanical vote counter, the phonograph, and, of course, the first commercial electric lightbulb. In the course of doing all of this, Edison had also created the first modern industrial research laboratory, using mass-production techniques that wouldn't be duplicated by most manufacturing companies for another decade.

The world had seen nothing like young Tom Edison, and it really hasn't since. There is no need here to recapitulate at length the story of Edison's early years—pulled out of school for lack of attention, getting a job as a telegraph operator after saving the life of the station-master's son, losing his hearing from either scarlet fever or getting his ears boxed by an angry conductor after Edison's secret lab set fire to a railroad car, learning about electricity from telegraph wires, getting fired, and living and experimenting in the basement of a fellow inventor—as most schoolchildren even today encounter the tale at some point.

If Edison's reputation today isn't as brilliant as it was a century ago, it's no doubt because its glow has been at least partially eclipsed by the growing exposure of Edison's sharp and sometimes questionable business practices. Today it is hard to get past the stories of ripping off Tesla, taking credit for the ideas of others, driving pioneering filmmaker Georges Méliès into bankruptcy, choosing *Birth of a Nation* as his favorite movie, and so on, in order to give Edison anything more than conditional credit for his accomplishments.

But dismissing him, even partially, would be a mistake. Because if it is memory that makes us human, and if memory is indeed the guardian of all things, the preserver of culture and wisdom, then no person in history made a greater contribution to the power, the range, and the democracy of memory than Thomas Alva Edison.

It began in 1869, with Edison's development of the modern stock ticker. Stock ticker tape machines (their name came from the sound they made and the paper tape on which they printed) had been around for more than a decade at this point, and their ability to send breaking financial data over long distances was instantly hailed as a vast improvement over human runners. But the early machines were big, clumsy, and unreliable. Worse, they printed out their streaming data in the form of Morse-like code that had to be translated. Edison's Universal Stock Ticker, by comparison, was a revelation. Almost small enough to fit in one's hand (it was typically, and memorably, covered by a glass dome), it was able to print on its paper tape actual letters and numbers. It was, in many respects, the first real-time capture of the memory of distant physical events at the instant they occurred.

But that was only the start. Shouting "Mary had a little lamb!" into the bell horn of his first phonograph in 1878, Edison captured not just the first human voice but the first recorded sound of *anything* on Earth. The phonograph's subsequent effect on human memory is almost incalculable, and we are haunted by the loss of the chance to hear the legendary voices and sounds we narrowly missed: Paganini's violin, a Dickens lecture, Edmund Kean's Shakespeare, Lincoln's Gettysburg Address. By the same token, we cherish those brief audio records of Enrico Caruso, Queen Victoria, Alfred, Lord Tennyson, and Florence Nightingale especially because we know that we will never hear the voices of nearly all of the other great figures of history.

Edison's original phonograph was a fragile device: The recording medium—a sheet of tinfoil wrapped around a cardboard tube—was only good for a few repeated plays, but that was enough to set the stage. In a way that we—after more than a century of continuous technological innovation—can barely imagine, the appearance of the Edison phonograph was like some kind of magic trick or something from another planet. Most people had never considered that sound might be a physical phenomenon, a wave of energy that could not only be transmitted—the telephone had been invented only two years before—but that might even, somehow, be captured and stored forever.

Edison's genius—echoed a decade later by Hollerith—was in understanding that this could be a reversible process of both recording and playback, and bringing together different technologies (horn speakers, reduction gearing, Bell's telephone microphone) to realize his goal. Artificial memory, which had been a solely visual process since the invention of writing, now, after five thousand years, added a second sense.

Within a few years, other inventors, including Bell himself, had improved upon Edison's design, notably by improving the recording technology and promoting a more durable wax cylinder licensed from Edison himself. By the end of the nineteenth century, German American inventor Emile Berliner further advanced the technology by creating the "gramophone"—the tone-arm needle/spinning-flat-disk design that would dominate sound recording for most of the twentieth century.

Edison stayed involved with sound recording into the early years of the new century, perfecting new, and more durable, recording media (including Amberol, an early plastic), but as was typical he grew increasingly cranky and resistant to change. Edison Records, which had once ruled the audio-recording world, finally made the transition to disk records in 1912—with a superior but proprietary technology that was not only more expensive but only played on Edison gramophones. Add to that Edison's unwillingness to pay top-flight talent to record for him (as did competitors like Columbia Records) and, on the eve of the Great Depression, Edison Records went out of business.

TUBE STAKE

By the early years of the twentieth century, recording sound and playing sound had divided into two increasingly distinct industries. On one side was the brand-new world of consumer entertainment—millions of gramophones produced by scores of manufacturers supported by growing catalogs of records sold by the tens of millions through an infrastructure of distributors and retailers, from record shops to department stores. On the other was the new entertainment industry providing the talent and content for these records; these included record companies, recording studios, and manufacturers.

It was on this latter side, content, where the important innovation continued to take place. Early recording sessions literally required performers to shout or sing into giant versions of gramophone horns in order to produce enough sound vibration to carve into the surface of a master metal platter that could then be mass-copied onto consumer records (in the early cylinder days, when only a dozen could be produced at a time, performers had to repeat their performances many times over the course of hours). Because the process—at both ends of the creator-consumer relationship—was mechanical, volume and sometimes fidelity were also limited.

That changed with the work of a whole new kind of inventor, one whose flamboyance and risk-taking style presaged wholly new generations of entrepreneurs to come. His name was Lee De Forest, and he was yet another Midwesterner (Iowa) who had gone east to make his name—in De Forest's case, via Alabama, where his father had been the controversial president of a black college. After earning his Ph.D. from Yale, De Forest took a faculty position at the future Illinois Institute of Technology teaching radio technology.

But De Forest didn't have the personality for a quiet career lived in a faculty lounge. And by 1905 he embarked on his own private research quest. He began with a device derived from Edison's lightbulb, but instead of installing a pair of electrodes with a carbon filament stretched between them, De Forest changed the location of those electrodes, making one end a filament and the other a receiving plate, which allowed the electrons to shoot across the gap through the vacuum. He filed the patent for this "diode," which detected electromagnetic waves, in 1906.

A year later, in the greatest innovation of his nearly two hundred patents, De Forest added a third electrode to this design—a grid, placed between the other two. The resulting "triode" is considered the first true vacuum tube . . . and even De Forest at first didn't understand what it was good for.

In the meantime, De Forest built a company to exploit what he believed would be a huge market for his two-electrode "Audion," as he called it, in the telegraph industry—where it proved useful in detecting wireless telegraph signals. But De Forest was too mercurial to be a good entrepreneur, and between his poor business skills and some crooked partners, by 1911, the company was out of business. De Forest moved to Palo Alto, California, and took a job with Federal Electric, one of the first technology companies in what would one day be Silicon Valley. But De Forest couldn't put his past behind him, and in March 1912, he was arrested by federal officers for stock fraud. It took a hurried collection of $10,000 bail by Federal Electric's directors to keep De Forest out of jail.

As he awaited his trial, De Forest embarked on one of the most feverish periods of creation in his life. And it was during one of these sessions, working with two assistants to try to perfect a newer Audion design, that

De Forest tried reconfiguring the locations of the three electrodes. To see how well it worked, De Forest plugged the tube into a telephone transmitter, put on a pair of headphones, dangled his "trusty Ingersoll" pocket watch in front of the transmitter . . . and nearly blew out his eardrums.[10]

Lee De Forest had invented the amplifier. And in the years to come, he and his successors would continue to experiment with the location of the components, the shape of the bulb, the transmitter "guns" shooting the electrons, and the coatings on the end of the bulb that would glow when hit by these electrons. The resulting inventions built from these new designs over the next forty years included radio, the klystron tube, radar, microwave transmitters, and television.

But one of the first places the tube amplifier found a home was in the recording studio—in particular, in microphones. The Audion brought electricity into the recording studio. Now sound could be boosted even as it was recorded, and that—combined with new master-recording technologies—meant that a single superior recording could be reproduced millions of times, and the master copy permanently stored. The performers could also hear themselves in playback over the studio's speakers. And once tube prices fell, that same combination of electricity and amplification could come to the gramophone—now called the "record player"—as well. Audio memory, in the form of what would eventually be billions of copies of records, LPs, cassettes, CD, and MP3 files, had permanently entered into the human legacy.

ENDURING IMAGE

Edison's final contribution to human memory took a little longer to create.

As already noted, he spent much of the 1880s dealing with both business and personal trials. But by 1888, he had a good idea of what he wanted to do next. He had remarried (to Mina Miller, the twenty-year-old daughter of another inventor) and moved from Menlo Park to homes in West Orange, New Jersey, and Fort Myers, Florida. He did little hands-on inventing anymore, but rather served more as an impresario of new ideas.

Edison's most famous quote had been that "genius is one percent inspiration and ninety-nine percent perspiration," but now most of that sweating was being done by an army of assistants, many of whom were better scientists and engineers than Edison had ever been. They were also more efficient: Whereas in the development of the lightbulb Edison had famously tried thousands of different materials before tripping over carbon filament, these new researchers had the skills to deal with the increasingly difficult technologies in a more theoretical and less plodding trial-and-error manner that had been Edison's forte. As the bitter Nikola Tesla would notoriously say in Edison's only negative obituary "tribute":

> *His method was inefficient in the extreme, for an immense ground had to be covered to get anything at all unless blind chance intervened and, at first, I was almost a sorry witness of his doings, knowing that just a little theory and calculation would have saved him 90% of the labour. But he had a veritable contempt for book learning and mathematical knowledge, trusting himself entirely to his inventor's instinct and practical American sense.*[11]

Of course, that didn't prevent Tesla from proudly accepting the Edison Medal a few years later.

But what this undoubtedly accurate observation missed was that Edison didn't have to be both the visionary and the builder to be a great inventor. That was a myth created by mid-Victorian inventors like Bell, Morse, and the young Edison himself, in an era when a visionary could also be a talented prototype builder. It wasn't true for very long, but that particular myth has had a very long tail: Even in the early twenty-first century, most of the general public credits Steve Jobs as the inventor of Apple's extraordinary run of consumer products, which wasn't true in fact, but was in the larger sense quite accurate.

Also the case in 1890 and in the years beyond was that the Edison Company became *more* effective the less the founder was involved in the detailed activities—from R&D to marketing and sales—of the company. A classic example of this was Edison's third great memory invention: motion pictures.

At this point, still photography—another great landmark in the history of memory—had been around for a half-century, arriving in time to give us indelible images of the Crimean War, Indian maharajahs, Chinese farmers, African tribesmen, and Mathew Brady's pictures of Antietam and Abraham Lincoln. Photography had been invented in the 1820s, but its core technologies (light-sensitive chemicals, lenses and cameras obscura, mechanical shutters) were already well established, with some, such as the pinhole camera, dating as far back as the ancient Greeks and the Chinese.

In the end though, it was Joseph Nicephore Niepce, a French inventor, who in 1826 coated a pewter plate with bitumen, exposed it for a number of hours to light to harden the coating, then washed away the unhardened portions with a solvent . . . and created *View from the Window at Le Gras*—a jumble of rooftops and towers that is history's first photograph. With it, he recorded the first image of the natural world that wasn't created by the esthetic intervention of man. Time would show that even photographs weren't bias-free, but they remain in daily life as close to an objective image of reality as we are likely to find. Artificial memory had now added its most important visual dimension.

Niepce, a remarkable inventor (he also built the first internal combustion engine), died in 1833, but before he did, he gave all of his notes on photography to his younger business partner, Louis-Jacques-Mandé Daguerre, a skilled theater designer who had already invented the diorama. Daguerre was a rare case of a skilled inventor who was also a good businessman, and in 1839 after years of improving the process, he filed for a patent and sold the technology to the French government. The resulting "Daguerreotype" was the first widely used photographic technology; and the name has been casually attached to every type of early photograph ever since.

However, great inventions almost never occur alone, and neither do they stay a monopoly for long. Across the Channel, William Fox Talbot was doing similar experiments on his "calotype" process. His greatest contribution to the story of photography was the development of the negative, which allowed multiple prints to be made from the original photograph—impossible with other contemporary photographic technologies. Meanwhile, over the next two decades a number of new and competing (price, quality, durability) technologies were developed, including "collodion"—the one used by Lewis Carroll for his Alice photographs—and tintypes, made famous by thousands of Civil War portraits, which were of inferior quality but offered the advantage to the middle class of being very cheap.

The limitations to these photographic technologies are quite obvious today as we look at surviving examples: They are dark, very fragile (the glass plates break, the images smear) and, because they required their subjects to stay very still for long periods of time, almost always stiff. That's the main reason why there are few "action" photographs from the Crimean and Civil Wars but a whole lot of images of the battlefield dead.

The biggest breakthrough in nineteenth-century photography took place almost exactly the moment that Herman Hollerith was designing his tabulator and Edison was working on his second-generation gramophone cylinders. The inventor-entrepreneur this time was George Eastman, a self-educated New Yorker who was one of the sanest of his breed.

Throughout the 1880s, Eastman systematically transformed almost every part of the photography industry—giving it the form it still exhibits today. He began in 1884 by applying for a series of patents for paper-based photographic film. Eastman had discovered a means to coat paper with a dry, light-sensitive gel . . . then had the brilliance to break from the existing plate paradigm and instead cut these sheets into long strips that could then be rolled. Next, he developed a camera that would accept these rolls, which he introduced to the world four years later. He then followed that, in 1892, with the founding of the Eastman Kodak Company ("Kodak" was one of the first made-up brand names) and set out to mass produce the world's first consumer camera.

This Kodak "box" camera was a huge hit, putting photography into the hands of the average person for the first time. But it was merely a prelude to Eastman's greatest success, the "Brownie" camera, introduced in 1901.

The Brownie, little more than a cardboard box with a lens in front, a button on top, and a roll of film inside, is a good candidate for the single-most influential consumer product model in history. Driven by its ad message, "You Push the Button, We Do the Rest," it sold in the tens of millions, and its descendants—built of Bakelite and armed with a flash and color film—were still the world's most popular cameras well into the 1960s.

FLICKERING FUTURE

As with the phonograph, the influence of paper photography and simple, low-cost cameras on the story of human memory is almost incalculable. Everyday Americans, and soon populations around the world, made the act of taking "snapshots" a regular part of their daily lives. And because they didn't have to deal with the development of the images but merely sent the Brownie to Eastman Kodak by mail and got back the printed photographs and a reloaded camera, owners took photographs by the hundreds: family portraits, vacations, work, novelty shots, public events, distant lands, famous figures. Together, all of these photographs constituted the greatest visual memory record of a culture ever accumulated to that date. Thanks to those giant archives of Brownie photographs we know more about the daily life of the average laborer at the turn of the twentieth century than we do about some of the more private monarchs living at the same time.

Needless to say, all of this made George Eastman a very wealthy man, which he immediately converted to equally impressive feats of philanthropy. He was the very model of the enlightened tycoon. In 1932, at age seventy-seven, after two years in agony from a degenerative spine disease, Eastman committed suicide. His last note perfectly captured the restless spirit of Eastman and his fellow entrepreneurs: "My work is done. Why wait?"[12]

The importance—and the implications—of Eastman's new film technology was recognized almost from the moment of its introduction, and even more so in 1889, when he perfected the process on transparent celluloid film. One person who fully appreciated it was Thomas Edison. Just a year earlier, Edison had attended a lecture by Eadweard Muybridge, Stanford University's master of stop-motion photography, and had been entranced by Muybridge's "zoopraxiscope," a disk with consecutive stop-motion images around its perimeter that, when spun, created the sensation of movement. Two days later, Edison met with Muybridge to discuss the technology, and the latter proposed a joint project that would combine the zoopraxiscope with the Edison phonograph. Edison declined.

But six months later, Edison, back to his old competitive self, notified the Patent Office of his plans to build a device to do "for the Eye what the phonograph does for the Ear." In his next caveat to the Patent Office, Edison gave this device a name: the *kinetoscope.* He then turned to one of his most talented assistants (and official company photographer), William Dickson, and gave him the assignment to start building a kinetoscope design that used tiny (1/32-inch wide) photographs developed directly on the surface of a rotating drum . . . and then to synchronize the rotation of this drum with the cranking of an on-board phonograph. Meanwhile, Edison took off for France, ostensibly to attend the 1889 Exposition Universelle in Paris, where the company had an exhibit, but also to investigate the current state-of-the-art technology in Europe.

During his two-month stay, Edison visited the naturalist-inventor Étienne-Jules Marey, who had invented the first motion-picture camera—a device that resembled a machine gun with a drum magazine—with which Marey exposed multiple images of the same creature on a single strip of flexible film. Edison also saw several other evocative new technologies, including Ottomar Anschutz's "electrical tachyscope," which used flickering light to trick the eye's persistence of vision to create the effect of motion; and Charles-Émile Reynaud's "Theatre Optique," which used perforations and a toothed wheel to pull handpainted film cells through a projector.

Meanwhile, back at Edison Labs, Dickson and his assistants had abandoned the drum model as too crude in its imagery and instead replaced it with strips of flexible film bearing consecutive images that were wrapped around the drum. In the process, they created the first-known moving picture on film made in the United States: *Monkeyshines No. 1* (1889). It lasted only a few seconds and featured a silly demonstration of physical dexterity by a fellow employee . . . but the door was now open.

When Edison returned from Europe, he quickly filed a new caveat with the Patent Office and then gave Dickson a new set of marching orders. Now the kinetoscope would abandon the drum and instead create a closed loop of film to be strung back and forth on rollers to tuck the maximum length inside the projector and pulled along at a steady rate in front of a projector at the right speed between frames to create in the eye and brain the sensation of continuous, smooth movement.

The design was sound, but Dickson and his crew were still having trouble with the film medium itself. They were reduced to hand-cutting sheets of brittle and stiff celluloid and gluing them together end to end. It was a wholly unsatisfactory solution. Then, in August 1889, while Edison was still in Europe, Dickson happened to attend a presentation by George Eastman of his new flexible, rolled photographic film—and knew he had his answer.

Then, remarkably, the whole project was put on hold as Edison pursued a mining project. It was 1891 before Edison and the team, now led by William Heise, set out to finish the design. The result, built within a large cabinet, was the loop of film, strung between the multiple rollers and then passed at the top through a viewing lens. The film itself was backlit by both a projector and a slit wheel that acted like a shutter to make the light flicker for a fraction of a second through the film to trick the eye. The crucial breakthrough proved to be an escapement mechanism, derived from watchmaking, which clicked the film forward in a unique stop-and-go motion every 1/46 seconds (in other words, forty-six frames per second) and would dominate motion-picture camera design for generations.

The first public demonstration of the kinetoscope took place on May 20, 1891, at Edison's laboratory for a group of 150 members of the National Federation of Women's Clubs. Wrote the *New York Sun*:

> *In the top of the box was a hole perhaps an inch in diameter. As they looked through the hole they saw the picture of a man. It was a most marvelous picture. It bowed and smiled and waved its hands and took off its hat with the most perfect naturalness and grace. Every motion was perfect. . . .* [13]

Edison spent the next five years rolling out the kinetoscope in galleries across the United States, improving the design by adjusting the film speed to improve total run time to about thirty seconds, setting up the Black Maria film studio to produce more content (most notably *Fred Ott's Sneeze*, the first copyrighted motion picture, and *The Great Train Robbery*, cinema's first classic), and even experimenting with sound. The biggest unveiling took place at the Chicago World's Fair, where the unexpected new miracle from the Wizard of Menlo Park swept away the crowd (as did the filmed belly dance of Little Egypt). Soon there were kinetoscope parlors from coast to coast—and a number of pirate companies creating and selling their own films for the machines.

During all of this, Edison went back to his original idea of adding sound to his kinetoscopes, calling them kinetophones. But his solution, to put a phonograph in the bottom of the case playing nonsynchronized sound, met with limited interest and was abandoned.

LAST GASP

In 1904 the kinetoscope business exploded in the United States, thanks to the number of installed units reaching critical mass; the creation of a new genre of films, prizefights; and the scandal (and arrests) surrounding another film—this one of a sexy dance by the music-hall performer Carmencita.

That same year, Edison opened kinetoscope parlors in Paris and London. In a move that is still debated, he chose *not* to file European patents beforehand.[14] One theory is that the notoriously cheap Edison refused to pay the $150 filing fee. Another is that he had violated so many European patents in designing the kinetoscope that he had no hope of ever obtaining such a patent. Either way, it was a dangerous move, and in time Edison would pay heavily for it. Not only did various entrepreneurs across Europe quickly steal his design (and revenues) but, perhaps worse, smart inventors "reverse-engineered" the kinetoscope, saw its weaknesses, and rushed to make competitive improvements.

Of the latter, the most important were the Lumière brothers, Auguste and Louis. Before making their place in history by filming (and permanently memorializing the memory of) everyday life in turn-of-the-century France, they permanently transformed the motion-picture experience by creating *projecting* kinetoscopes that showed their films on screens in front of audiences.

By 1910 the Edison Company was already an also-ran in the movie business. It continued to produce innovations, such as a truly synchronized kinetophone in which the (now projecting) kinetoscope was linked to the phonograph via a belt. But it was hard to maintain in country theaters and didn't sell well. Another interesting attempt—a home kinetoscope—failed as well, largely because it was so far ahead of its time. Finally, a fire at Edison in December 1914 put the company out of the movie business forever.

But by then, Thomas Edison had made his final great contribution to human memory. His early films represented the first time that the world as it was actually lived, in full motion, was captured and preserved forever. A century later, as we look at these early films, as well as those of the Lumières and others, we often find ourselves staring past the foreground performance to get a glimpse of a larger world now lost forever and to see ourselves in people long since departed.

Thomas Edison died on October 18, 1931. The world mourned. Henry Ford asked Edison's son Charles to capture and seal a test tube containing Edison's last breaths. Two years earlier, on the fiftieth anniversary of the incandescent lightbulb, the world shut off its lights to remind itself of what life had once been like. But when the lights went back on and the populations turned to their record players or went to see a movie, they equally honored Edison, for he had given them—and us—the memories of our present into the endless future.

SOUND AND FURY

Thomas Edison may have failed to effectively synchronize sound with film, but there was another genius-inventor waiting in the wings: the ever-mercurial Lee De Forest.

Having transformed music recording, sound reproduction, and eventually radio and telegraphy with his audio tubes, De Forest now set about bringing sound to silent films. He was convinced that there must be a way to put the audio directly *on* the film, rather than the primitive technique of dropping a needle on a record timed with the start of a motion picture.

This was 1918, and audiences were still getting accustomed to the thrill of watching multireel movie extravaganzas such as *Birth of a Nation* and *Intolerance*, in which a whole vocabulary of acting and filming had adapted to the lack of a soundtrack. These audiences hadn't evinced much interest in sound movies. But De Forest was undeterred. And in 1919, working from basic research by the Finnish inventor Eric Tigerstedt, De Forest patented the first sound-on-film technology, which he called "Phonofilm."

Phonofilm in action was simplicity itself, and a testament to De Forest's brilliance. Rather than try to meld the phonograph and film, De Forest instead decided to "film" sound—that is, he used a narrow strip on the side of the roll of film to capture the image of the sound waves picked up by microphones in the filming process. The sound waves were presented and saved as narrow lines of different gradations of gray. Then, as the film ran through the projector, these millions of parallel lines were converted back into their corresponding sounds and projected over speakers.

De Forest premiered Phonofilm in 1922 with a collection of short movies of speeches, stage performances, and musical acts . . . and waited for the movie industry to shower him with money. Instead, the movie studios ignored him. The already-paranoid De Forest assumed—this time perhaps correctly—that there was a conspiracy against him. A more likely explanation was that the studios had enough on their plates just getting films out with the current silent technology and appreciated more than De Forest the industrywide chaos that sound would (and did) create. And besides that, they probably didn't want to pay De Forest's licensing fees and planned to hold him at arm's length until they could come up with a similar technology of their own.

De Forest refused to give up, and began producing short sound films of his own. He even convinced the Fleischer brothers to create a series of "follow the bouncing ball" sing-along cartoons that were still being shown on television into the 1960s. But the irascible De Forest, who tried to take credit for everything, eventually drove away his business partners, who turned around and sold his patents to Fox Films.

The year 1927 saw the birth of the "talkies"—many of them using Phonofilm audio technology. But De Forest wasn't there to celebrate. Having been married and divorced three times already (once to a famous suffragette), De Forest married and settled down with a movie ingénue from the Hal Roach Studios, Marie Mosquini, and proceeded to publicly decry the debased uses of his invention—most famously in an open letter that announced: "What have you done with my child, the radio broadcast? You have debased this child, dressed him in rags of ragtime, tatters of jive and boogie-woogie."

He lived long enough to earn an Academy Award for his work, and a star on the Hollywood Walk of Fame.

TALE OF THE TAPE

Alexander Poniatoff always thought big, even when he was thinking small.

Born in tsarist Russia at about the same moment that Thomas Edison was showing off the kinetoscope in the United States, Poniatoff dreamed of growing up to design and build great locomotives. His father was a wealthy lumberman in Kazan, and so when young Alexi came of age, he was sent off to a technical academy in Karlsruhe, Germany, to study engineering.[15] By this stage his dream was to return to Mother Russia and build a great turbine factory.

But it wasn't meant to be. Alexi returned to a Russia mobilizing for war. Then came the revolution . . . and then the civil war. A White, Poniatoff joined the army and trained to become a pilot. But as defeats mounted and the Bolshevik Reds consolidated control, Alexi realized that his cause was doomed. He deserted and made his way to China. He found work at the Shanghai Power Company.

Needless to say, China wasn't the best place to be in the 1920s as it slid into its own revolution. So, in 1927, Poniatoff emigrated again, this time to the United States. By now one of the few experienced electrical engineers in the country, Poniatoff found himself recruited by a number of companies, including Pacific Gas & Electric, Dalmo-Victor, and Edison's old company, the now-giant General Electric.

Poniatoff's skills were in even more demand when World War II erupted, and he spent most of the war years designing motors and generators for airborne radar systems. This classified work, combined with Poniatoff's highly respected expertise, enabled him to be one of the first scientists to get a glimpse of the secret technological spoils emerging out of the fallen Third Reich. What he saw made Alexander Poniatoff quickly abandon all of his own dreams in pursuit of a new one.

Magnetic recording, though it seemed revolutionary when it first appeared, had already been around for a very long time. In 1888, Oberlin Smith, an American machinist, devised a system for attaching steel dust to a fine thread. He then pulled the thread past a magnet that was in turn attached to a microphone. This process magnetized spots on the thread in relation to the transmitted sound signal. But Smith left it at that, published his results, and went back to making machine tools.

A decade later, Danish engineer Valdemar Poulsen picked up on Smith's theories and ran with them. He used wire this time (which Smith thought couldn't be done), which he wrapped around a drum, and improved the recording/playback magnet ("head"). He called the resulting design a "telegraphone." Because there was no amplification, the recordings were weak, but if one used headphones, still hearable. Poulsen showed his telegraphone at the 1900 World's Fair in Paris, and the recording he made there of Emperor Franz Josef of Austria survives as the world's oldest magnetic sound recording.[16]

Amplifying the signal of a wire recorder was a lot more difficult than it seemed. Merely boosting the power of the signal resulted in considerable loss of the lower frequencies and a lot of distortion at higher frequencies. Moreover, adding this power, especially with direct current, also overly magnetized the head, creating further problems. Finally, the wire medium, reduced to the thickness of a human hair to get the mile-plus length on the spool to create an hour-long recording, had a tendency to twist and tangle during rewind and editing. Still, it was better than the alternative of having to work with carved master disks; to make a splice with wire the editor merely held his cigarette to the two ends and welded them together.

The problem of amplification was eventually solved through a process called "bias," by which a controlled AC signal was added in a specific pattern to the signal before and after the read-write head, thus removing any existing magnetism while moving the recorded sound into a better range and then boosting the resulting signal.[17]

By the late 1920s, magnetic wire recording had been improved to the point where it could be used for office dictation and telephone recording, and a number of companies sprang up to pursue those opportunities. But wire recording remained generally unknown to the general public, which might have been surprised to learn that several popular radio shows, the first being Edward R. Murrow's *See It Now* on CBS, were recorded and edited on wire.[18]

In World War II, both sides made heavy use of wire-recording technology. One of the most interesting applications was that of the "Ghost Army," in which the U.S. Army Signal Corps took wire recorders to the front lines to play military sounds in order to confuse the enemy. Immediately after the war, psychology professor David Boder rushed to Europe with a wire recorder and conducted numerous interviews with Holocaust survivors—among the most historically important recorded memories of the twentieth century.

As the war neared its end, rumors began to circulate that the Germans had developed a major improvement on wire recording. And indeed they had. Beginning in the late 1920s researchers had begun to look at ways of improving the wire medium in order to capture more of the analog data emerging from the source. The obvious solution was to make wire "wider"—that is, to turn it into a metal strip. Just such a steel tape recorder was first used by the BBC in 1932. But the technology proved not only unwieldy—a one-hour taping required 3 kilometers of tape racing past the read-write head at 1.5 meters per second—but hugely dangerous: If the spring-steel, razor-sharp tape were to snap, it would thrash around the studio, slashing everything in its path.

But as war clouds gathered, a group of scientists at the IG Farben chemical company subsidiary BASF, working with the Third Reich's official propaganda radio network, revisited magnetic-tape recording and came up with the first practical solution: plastic tape, coated with iron oxide and run through a ring-shaped head (less destructive than the traditional needle head) and amplified using AC bias. The results were stunning—and not lost on the Allies, who raced to figure out how the Nazis were able to repeat broadcasts to different time zones.

When the Allied scientists, including Poniatoff, finally got a chance to see the German tape recorders, they knew they were seeing the future . . . and it wasn't long before companies in the United States and Western Europe initiated their own magnetic tape-recorder development programs.

One of these competitors was a brand-new company founded by Alexander Poniatoff. He called it *Ampex*, after his own three initials, followed by "EX-cellence."

What Poniatoff, with his peerless experience in electrical engineering, saw about these recorders that was missed by many of his competitors was that the fundamental challenge was not in making the technology work better, but in making the practical recording time *longer.* That seemed impossible: The tape was already flying through the recording head, and it couldn't be made much thinner, so the only solution seemed to be bigger and bigger reels.

But Poniatoff had a better idea: Instead of making the tape go faster, why not slow it down to add recording time . . . and make up the difference by spinning the recording head instead?

Poniatoff wasn't the only entrepreneur thinking outside of the box when it came to recording technology. Major Jack Mullin was in the U.S. Army Signal Corps, assigned to find out everything he could about German electronics and radio technology. By chance, in 1945, just as he was heading home to California, he stopped to inspect a newly captured radio station in Bad Nauheim, near Frankfurt. There, he found two suitcase-sized "Magnetophon" tape recorders and fifty reels of Farben tape. He shipped them home to spend some time working with them.

In 1946, after a demonstration for engineers in San Francisco of his improved Magnetophons that met with an enthusiastic response, Mullin decided it was time to pitch his recording technology to Hollywood. His timing was impeccable.

The biggest recording star in the world in those days was Bing Crosby. Crosby, who preferred the casual intimacy of the recording studio to the stopwatch world of live radio, had been fighting MGM for the right to record his radio show. Citing poor recorded sound quality, MGM had refused, and a small war of nerves had erupted—to the point where Crosby had even briefly quit radio in 1946. So, when Mullin demonstrated his Magnetophon at MGM one afternoon, Crosby's technical director Murdo MacKenzie knew he had heard the answer. He quickly arranged for a meeting between Mullin and his boss.

Crosby, too, was impressed with Mullin and his machine—and quickly wrote out a $50,000 check to cover the purchase of the machine and to make an investment in its manufacturer. Mullin, however, didn't have a company—but he knew who did: Alexander Poniatoff and his six-man Ampex, where Mullin was a consultant. Poniatoff, who had just completed his own design for a rotating-head recorder, the model 200, filled the order. Crosby and his team went on to use the editing features of the Ampex 200, and 3M Company's new acetate magnetic tape, to revolutionize radio broadcasting (including the notorious laugh track), Mullin got very rich, and Ampex became the fastest-growing company in business history—a pace that wouldn't be equaled until the dot-com boom of the 1990s.

In little more than a decade, Ampex grew to thirteen thousand employees and utter dominance of the audio-recording industry. Guitarist Les Paul used an early Ampex recorder to edit together multiple recordings into one—the beginning of multitrack recording. Elvis Presley would make his first recordings at Sun Studios on an Ampex reel-to-reel. Elizabeth Taylor's husband, Mike Todd, worked with Ampex to place a magnetic strip on film to carry a much higher level of audio quality in movies. And having captured almost the entire professional recording world, Ampex began building consumer-grade tape recorders in the late 1950s, capturing that market as well.

By the mid-1950s, Bing Crosby was experimenting with ways to record video signals on tape as well. Once again, Poniatoff took this idea and ran with it, assigning a team that included nineteen-year-old future sound wizard Ray Dolby to build it. The team came up with a design that ran two-inch-wide tape at fifteen inches per second across four heads that were spinning at almost 15,000 rpm. The first taped network television broadcast, the *CBS Evening News,* was broadcast on November 30, 1956. Within thirty years, 100 million home videocassette players would be in use around the world, showing billions of professional films and homemade videos created on a new generation of handheld video cameras. The Kennedy assassination would be captured on videotape. So would man's first step on the moon. Video memory, the defining medium of artificial memory in our time, was born.

ERASURE

There remained one last great market for magnetic tape. Computers, the descendants of Herman Hollerith's tabulator, had evolved slowly during the early years of the twentieth century, serving as little more than sophisticated calculators. But in the 1930s, once again driven by the oncoming war and the need for powerful tools for everything from encryption/decryption of codes to the computation of artillery trajectories, computer technology developed at a rapid pace throughout the world.

In the UK, the great mathematician Alan Turing built a series of computers that used the ones and zeros of Boolean algebra to create increasingly powerful (and eventually tube-driven) computation engines for code-breaking. In Germany, Conrad Zuse used electromechanical relays in his Z series—making them the first electronic computers—to compute artillery tables at record speed. But it was an American, Claude Shannon, who put together the two technologies—Boolean logic and electrical relays—to devise the modern computer architecture. It would be realized in Harvard's Mark 1 and, just after the war, the ENIAC.

The fast-moving, fast-growing, information-driven postwar era was just made for the computer, which in turn made that world possible. Many companies fought for the military and then commercial computer market, but one company emerged on top: IBM, the company Thomas Watson Sr. had built from Hollerith's struggling start-up.

These early "mainframe" computers were the size of a small building and glacially slow by modern standards, but they were still fast enough to quickly outstrip the devices designed to put data into them and then take out the processing information they produced. Indeed, the first great postwar mainframes depended almost entirely on two nineteenth-century technologies: Hollerith's punched cards (and tapes) and Edison's alphanumeric printer.

It was the beginning of the still-ongoing race by artificial memory to keep up with the ever faster demands of the digital world. This time the answer came from the typewriter company Remington Rand with its UNIVAC computer line. In 1951 it licensed existing magnetic tape recording technology, adapted it for digital signals, and introduced UNISERVO—the first computer magnetic tape memory system. A year later, IBM introduced a seven-track tape memory and quickly ran away with the market by coming out with a series of milestone tape players over the next decade (and leveraging the power of its leadership in mainframe computers). By the 1960s, the room full of tape memory machines, each with its spinning spools, had become synonymous with the computer itself.

In time, magnetic tape memory was supplanted by other, more powerful digital memory technologies. Nevertheless, though rarely noticed, magnetic tape technology remains the world's most commonly used memory medium—in the form of the magnetic stripes on the back of credit cards. This format was invented in 1960 by IBM engineer Forrest Parry and proved to be an immediate success. It is estimated that in 2010 80 percent of the world's population used magnetic stripe technology in some form—and that these cards were swiped through readers 50 billion times per year.[19]

As for Ampex, the company seemed to grow old quickly, as if its health was tied to the aging Alexander Poniatoff. Poniatoff had been a middle-aged man when he founded Ampex, and in 1955, when he became chairman, he was sixty-three. Almost as if he was bored with his success, he began to take the company into ever more risky new ventures—notably a complete content-production operation on which the company lost a fortune. And he was cavalier about the company's assets: Contemptuous of that nation's cheap manufacturing, he casually licensed Ampex's recording technology to Japanese electronics companies . . . and then watched as those firms captured the entire consumer electronics industry and made that country rich.

When Ampex tried to fight back, it found that its long-standing philosophy, in the words of a former executive, that it "only knew how to do things well and costly," no longer worked. Its new home-video recorder was the best in the world . . . but also oversized and overpriced—and doomed.

So was Ampex. But Alexander Poniatoff barely seemed to notice. He was older than the century, the last tsarist—"A real eighteenth-century man," said one employee—and showed it with increasing eccentricity:

> *In the 1970s he became a health nut. He ate only unprocessed foods, drank carrot juice and had one of the first air ionization systems installed in his office. He began backing medical groups studying longevity and the effect of color on behavior. He drove only white cars and took to wearing a baseball cap. Mrs. Poniatoff held her Horticultural Society meetings in the Ampex cafeteria.*[20]

There were even rumors of séances. If so, they must have predicted a bleak future. After Poniatoff died in 1980, at age eighty-eight, Ampex prudently stripped itself of all of these extraneous ventures and went back to its core business of selling professional recording equipment. But it was too late.

Today, other than a skeleton team managing the company's once-great intellectual property assets, all that remains of Ampex, the company that did more than any other to capture the memory of the sound and look of this world (and others) is the huge old company sign, now a historic landmark, that towers over Bayshore Freeway in Silicon Valley. It stands in memory not only of one of the most remarkable companies ever but of the golden age of entrepreneur-inventors.

9
Diamonds and Rust

Memory as Free

Any one of the diminishing ranks of people who remembers daily life with audio tape (reel-to-reel, 8-track, cassette), videotape (videocassette), or computer memory tape knows well the shared limitations of those technologies.

Eight-track tapes had superb sound reproduction—better than today's MP3 files. Videocassettes did, too, and their fast-forward and reverse operations were more precise—and often easier—than today's DVDs. And computer memory tape reels held massive amounts of data—more than their successors could match for years.

But tape, in any form, had one gigantic, infuriating, and ultimately fatal limitation: *It was linear.* That is, all of the memory stored on magnetic tape was sequential: The next item to be stored was encoded right after the last one. And what this meant in practice was that the average time to find anything on a particular tape was one-half the length of that tape. That is, if you were lucky, the memory segment you were looking for was directly adjacent to where you were on the tape. At worst—usually when you were trying to set the mood on a date—it was at the other end of the tape. And heaven forbid if you initiated your search in the wrong direction. . . .

What made all of this particularly frustrating was that the world was still happily using a technology nearly a century old—the phonograph—in which it was a simple matter to locate a memory by merely picking up the needle and dropping it elsewhere on the surface of the record—a process that might take minutes cycling through a tape. It was the story of papyrus scrolls all over again.

Tape makers tried to overcome this inherent weakness by placing multiple tracks running in parallel on the tape, so that the operator could save time by jumping from one track to the next, but that also meant the information content (i.e., the width) of each of those tracks was now reduced.

Still, with some modifications in iron-oxide density, faster head speeds, better spindle motors, and denser tracks, the magnetic tape industry might have managed to keep up with the larger world of audio entertainment, television, and computers—had everything else remained intact. But those industries did nothing of the sort. On the contrary, these were some of the fastest-growing industries in business history.

The situation became especially acute in the case of computers (indeed, audio and video tape might have otherwise gone on for another generation as the medium of choice for consumers). Because the evolution of computers was taking place so quickly, and the amount of data that needed to be stored in memory was growing so exponentially, memory storage seemed about to be overwhelmed.

To understand why this was the case, we need to take a quick look at the history of computers to this point.

COLLECTING BITS

Early computers like ENIAC didn't need much memory. They were primarily performing discrete operations, such as large-scale computation. Their operators essentially entered the raw data using switches—or, when more throughput was required, they used the nineteenth-century technologies of paper tape, punched cards, or typewriters rewired to send signals. Output was largely the same thing, with the typewriters now converted to printers.

In the late 1940s, as computers became more sophisticated and began to assume more tasks in statistics, finance, and testing and measurement—that is, as the quantities of data going both in and out of computers grew—the architecture of these machines began to change as well.

Computer architecture has three basic components: input/output, which brings data into the computer and brings out results; logic, which is the central, computational part of the computer; and memory. In the early days of computing, with input and output comparatively simple and memory mostly taking the traditional form of "printouts" on paper or tape, much of the industry's concentration was focused on logic and the central processor. Because this operation was designed to be very fast, it was largely done with great banks of vacuum tubes. Tubes, essentially variations of De Forest's triode, were very hot and short-lived (the technicians for the ENIAC ran around in bathing suits *inside* the huge computer, changing burned-out tubes every few seconds), but they had no equal for speed. One of the first assignments given to ENIAC, for example, was to solve a problem related to nuclear physics. The problem, estimated to take one hundred scientists a year to answer, was solved by ENIAC in *two hours*.[1]

Computer memory, meanwhile, was secondary, and when tube memory proved too expensive and unreliable, computer companies just stuck with paper printouts to deal with what were becoming huge "batch" operations. That's why the invention of magnetic tape memory was welcomed by the industry. With tape, these giant new processing events—such as payroll—could be output onto big reels and saved for later printing as needed.

However, computer memory was not just growing in importance but also scope. By the 1960s, the industry had consolidated into what was known in the United States as "IBM and the Seven Dwarfs" (Burroughs, Univac, Control Data, NCR, Honeywell, General Electric and RCA) for their comparative size, as well as Olivetti and Siemens in Europe, and Hitachi, Fujitsu, and NEC in Japan, among others. This was the era of "Big Iron," as exemplified by the world-dominating IBM 360 Series. By now there were—if one looked at the data-processing world's entire data pathway—*six* different kinds of memory relating to computing.

First, there was all of the memory, in the form of raw data coming off test and measurement instruments in laboratories; payroll, tax, financial, and personnel records being created throughout organizations; and statistics, such as from the latest census, that had to be stored on forms, paperwork, and field notes before being entered into the computer.

Second, there was internal memory—called "random access" or RAM, because it was kept in an undifferentiated area inside the computer and given an address where it could be located.

Third, was "cache" memory. The central processor of computers worked at a very fast pace defined by the frequency of its internal signal—called its "clock speed." Because there was no way that memory could keep up with this pace, the computer would look ahead at operations to come and import the needed data into the cache—a kind of data waiting room—so it would be there when needed.

Fourth was "read-only" memory, or ROM. Computer users discovered early on that there were certain operations that were constantly in use, and that reloading them into the computer was a waste of time. The result was the creation of a region of memory designed to permanently hold these programs in a secure place in the computer where they could be easily accessed but not easily replaced or modified. Early computer programmers had also learned pretty quickly that, rather that writing all of the steps of a program in the most fundamental computer assembly code, they could develop "languages" (among the most famous being COBOL, created by future rear admiral Grace Hooper of the U.S. Navy) and tools to simplify their work and to create a sort of memory library of programs to build upon.

Fifth was archival memory. In a typical operation, the results from the computation process went back to RAM waiting to be downloaded as output. However, in the early days of computing there wasn't enough memory to go around—hence the appeal of tape memory as a place to store information peripherally to the operation of the computer itself. Today the last regular use of magnetic tape with computers is as long-term information backup and archive for personal-computer memory security.

Sixth—and often not counted in the process—was output memory. That was what happened to all of that processed output when it was eventually put to use in offices, laboratories, and classrooms. Despite all of the talk of the impending "paperless office," the real result of this explosion in computer output was mountains of paper printouts that were difficult to access and use.

A NEW SPIN ON MEMORY

That was a lot of memory—and not a lot of good solutions for managing it. By the 1950s, the need for reliable, fast, and electronic memory had become desperate. There were several candidates, and all were put to use.

One of these was magnetic *core* memory. This technology, invented separately by IBM's Frederick Viehe and Harvard's Way Dong-woo and An Wang (the last of whom later became a computer industry tycoon), consisted of tiny magnetic metal rings—the "core"—woven together on a grid of metal wires. Each of these rings, addressed through the warp and weft wires of the grid, could then be turned on and off (1 or 0) as a single bit of memory.

Core memory was not easy to manufacture but became cheaper after computer companies found low-pay workers around the world to do the tedious stringing—most famously, Scandinavian seamstresses who had been left unemployed by the automation of the local textile industry. Core memory was fast and had the unique advantage of retaining its data even after the power was turned off. But it was very difficult to test for failures, and most important—a weakness that eventually led to its demise—it didn't scale well. Every added bit meant an added ring—an unsustainable model when memory reached millions of bits.

The second solution had an unpromising start but proved to be both enduring and remarkably adaptable. Drum memory, originally designed for audio recording, was invented in 1932 by punched card–maker Austrian Gustav Tauschek. He likely saw it initially as an audio recording device, and shrewdly went back to the iconic drum-cylinder format of the early phonographs and kinetoscopes. When IBM bought his company, the patents went with it and became a long-term initiative in Big Blue's labs.

By the 1950s, the drum-memory design that emerged involved "painting" the outside wall of a spinning drum with ferromagnetic material—typically iron oxide (rust)—then stacking an array of read-write heads up against that wall. The drum was then spun rapidly and the heads created multiple tracks. This, of course, was very similar to magnetic tape; you still had to wait for the rotation of the surface to get to the right spot, but the sheer number of heads and tracks cut the search process considerably.

But at its best, drum memory barely matched the capacity of magnetic tape, and its access times were much slower than core memory. Still, it pointed the way to a far better solution. That project, too, began at IBM, but the work was taking place in San Jose, California. That location proved important because the creation of this new technology—*disk memory*—would require all of the creativity, improvisation, and contempt for rules for which Silicon Valley was already becoming known.

Rey Johnson hadn't planned to be one of the century's great contributors to artificial memory. He had, in fact, started out as a high school science teacher in Michigan. But he was a born problem solver, and when he designed an electromechanical device that could automatically "read" pencil marks on standardized multiple-choice test forms, he thought the maker of those forms, IBM, might be interested. But IBM wasn't interested. Johnson shrugged and went back to teaching.

That was 1932. Two years later, as standardized testing took off and began to swamp the small army of human graders, Big Blue took a second look. This time IBM offered Johnson a job as an engineer in its Columbia University and Endicott, New York, laboratories. There, Johnson spent twenty years becoming an acknowledged expert in punched card–memory printers and sorters.[2]

The year 1952 was a landmark year in the story of computers. As legend has it, a survey had estimated that the entire U.S. market for mainframe computers was seventeen machines. IBM's new CEO, Tom Watson Jr., son of the founder, took an enormous risk and ordered his company to build nineteen—the first to be delivered in 1952. By then, IBM knew it had made one of the smartest decisions in business history.

But now the company had a whole new set of problems. IBM alone was producing 16 billion punched cards per year—and customers were complaining not just of the storage problems for all of that thick stock paper, but that many of these stacks of cards had to be loaded every day for the same purpose, when in theory the same standard programs could somehow just be loaded into the computer and downloaded automatically as needed. Concerned about this dawdling pace of computer memory development, IBM sent Rey Johnson to San Jose to open a new company research laboratory. Johnson recalled:

> *I was told that my flair for innovative engineering was a major consideration in my selection to manage the new laboratory. During eighteen years with the IBM Endicott laboratory, I had had responsibility for numerous IBM products—test scoring, mark sensing, time-clock products, key punches, matrix and nonimpact printers, and random card file devices. By 1952, I held over fifty patents, some of them fairly good. To be given freedom to choose our projects and our staff made the San*

Jose laboratory an exciting opportunity, especially since funding was guaranteed—at least for a few years.[3]

Santa Clara Valley was mostly orchards in those days, and San Jose an agricultural town, but the company was shrewd enough to recognize that something important was taking place just below the area's bucolic surface. Stanford University was there. So was Hewlett-Packard. And the beginnings of a NASA facility at Moffett Naval Air Base. Ex-soldiers, who had seen the region on their way to the Pacific, were now moving west, armed with their GI Bills. And not least, the Lockheed brothers, having made their fortune in Burbank with airplanes, were now making plans to move back to their childhood home to build rockets and missiles. IBM's new San Jose laboratory, just outside of downtown in a warehouse near a popular barbecue joint, was but one of many new start-ups being set up along the new Bayshore Freeway south from Palo Alto.

Johnson's assignment was to gather a team and then investigate all likely new forms of high-volume, high-speed computer memory.

Naturally enough, he and his team began with drum memory, pursuing an alternative design of a rotating drum coated on the *inside*, in which a single read-write head mounted on an armature raced up and down across the tracks. But the results were unsatisfactory. So the team moved on to magnetic tape loops, magnetic plates, magnetic tape-strip bins, magnetic rods, and even revisited magnetic wire. In the end, the team settled on magnetic disks because of their high surface area, easy rotation, and multiple points of access.

Fortuitously, at almost that very moment a request for a bid came in from the U.S. Air Force Supply Depot in Ohio. The depot had been using a mainframe computer to manage its massive inventory, and despite the obvious advantages in using such a powerful machine to keep track of hundreds of thousands of items, the operators had become increasingly disappointed by the experience. The problem once again was lag time: Because of the "batch" nature of contemporary computing, huge lists of arriving and departing items piled up before the computer's records could be updated. This meant in practice that at any given moment the depot's records were now both more *and* less accurate than before.

What the depot wanted was a way to add and subtract inventory on its computer in "real time."

Johnson and his team, which was looking at similar applications at grocery stores in the Bay Area, believed they had the answer—and set about trying to build a prototype of this "disk" memory. Wrote pioneering tech journalist George Rostky, "When management back East got wind of the project, it sent stern warnings that [the disk project] be dropped because of budget difficulties. But the brass never quite caught up with the cowboys in San Jose."[4]

DUCK AND COVER

Even if the theory was good, the actual physical engineering of such a new technology was extremely complicated. The disks, which were aluminum and two feet in diameter, were heavy and they had to be perfectly centered on a tiny spindle. In the initial test the designers were prepared to duck if the spindle snapped and shot a homicidal metal Frisbee around the lab. But it held. Better yet, the entire test construct of 120 disks on a single shaft separated by quarter-inch spacers stayed intact even when rotated at 3,600 rpm.

Then there was the matter of getting the iron-oxide paint—the same as used on the Golden Gate Bridge—to smoothly and evenly coat the surfaces of the disks. One solution—to pour the paint in the center of each disk and let centrifugal force spread it out over the disk's surface, just like the paint wheels in carnivals—was tried and abandoned. In the end, another engineer on the team found the solution. He showed up at the lab one morning with one of his wife's silk stockings. Filling a paper cup with the precise amount of paint, he sprayed it evenly through the stocking . . . and achieved just the right thickness. A variation of that technique became the standard for the industry for years to come.

The final challenge was finding a way to make the read-write heads "float" over the surface of the disks close enough to read the magnetic record below without actually dragging across it and damaging the iron-oxide coating. It was a young UC Berkeley grad student, Al Hoagland, who came up with the solution of pumping air through nozzles in the read-write head to create an air cushion between the two surfaces.

On February 10, 1954, Rey Johnson and his team hooked a keypunch machine to the prototype and entered data onto the disk drive; then they reversed the process and had the same data printed out on punch cards. With characteristic plainspokenness, Johnson wrote into his lab notes: "This has been a day of solid achievement."

In early 1955, IBM ordered fourteen machines.

The resulting commercial product, the IBM 305 RAMAC, introduced in September 1956, was a milestone in the story of computing, helping to make possible the first real-time mainframe computers. In turn, in the decades to come, they would lead first to minicomputers and workstations, then personal computers and, in the form of servers, the Internet.

Compared to the modern hard-disk drive, which may stuff a trillion bits on a sliver of a disk the diameter of a silver dollar in a case the size of matchbook and weighing less than an ounce, the RAMAC was not just primitive but positively gargantuan. The size of a small closet, it contained fifty 24-inch-diameter disks and literally weighed a ton, had to be moved with a forklift, and was delivered via cargo planes. It held a total of 40 million bits. And it was puny compared to the 1961 Bryant Computer disk drive, which held twenty 39-inch platters producing so much centrifugal force that the system had to be bolted to the cement floor to keep it from "walking" across the room.

It also cost $150,000 in 1956 dollars. But for customers—from the first (Chrysler's MOPAR division), to nearly the last (the 1960 Winter Olympics)—RAMAC was worth every penny in the cost savings that came from processing turnaround times that had been improved by an order of magnitude.

Johnson went on to lead his team to further improvements of the RAMAC design. In the late 1960s, he entered the story of memory one more time: As a consultant for Sony Corporation he was asked to find a way to make video more available for schools and kids. Concluding that the problem was that Sony's one-inch videotape was too heavy and unwieldy for small hands, he took a spool of the tape, cut it in half lengthwise to a half-inch width, and then encased the tape in a plastic holder that made it accessible. And thus Rey Johnson, the inventor of the disk drive, also became the inventor of the videocassette.

MYSTERY DISK

By the early 1960s, IBM and competitors such as NCR were racing to create ever-smaller, faster, and denser disk systems. By now they were the size of dishwashers and even featured removable disk packs. And the influence of these new systems was profound. IBM's latest generations of disk drives were crucial to the operation of its new 360 Series mainframes, which, along with its successor the 370 Series, was arguably the most influential and dominant computer line in history.

But as in the early 1950s, IBM also knew that it was headed for a new development wall if it couldn't discover yet another quantum leap in memory design. Once again, the answer came from its San Jose laboratory, now under the direction of Ken Haughton. The plan for this new drive, ultimately called the Model 3340, was to create a configuration of two removable thirty-megabyte modules. History has often assumed that IBM gave the development of this drive the code name "Winchester" because of the nearby Winchester Mystery House, located on Winchester Boulevard in San Jose. But in fact it was Haughton who saw the planned configuration and said, referring to the famous rifle, "If it's a 30-30, then it must be a Winchester."[5] When the 3340 proved to be a technological breakthrough of historical importance, the code name stuck—and even today is used to describe hard-disk drives with similar technology.

The key to Winchester technology was the use of a very light read-write head that could be flicked back and forth across the tracks on a disk almost faster than the eye could see. And the key to doing this was the team's realization that if the head was sufficiently light and slightly curved like a wing there would be no need for the injected air; instead it would produce enough lift to "fly" 18 millionths of an inch above the surface of the disk (and land in specially designated landing zones on the disk). With Winchester technology, the Model 3340, introduced in March 1973 for the 370 Series, a user could find any record on the surface of the disk in no more than 25 milliseconds.[6]

Two years later, IBM advanced this technology still further by introducing "thin film" heads for these drives. This photolithographic technique put the entire wiring of the read-write on a tiny sheet of film, which allowed the flying head to be even lighter and smaller. The combination of Winchester technology with thin film heads was first introduced in the IBM 3380 of 1980.

By then, a whole new disk technology had emerged, most of it led by a "dirty dozen" (as they were called by IBMers) of memory specialists who walked out of the bureaucratic and conformist Big Blue to start their own companies.

The most important of these entrepreneurs was Alan Shugart Jr. Shugart had a long and successful career at IBM and had then jumped for a few years, from 1969 to 1973, to one of Ampex's biggest competitors, Memorex. Now, having gathered a team and found venture capital money, Shugart started his own company, which he called Shugart Associates—and set out to compete with IBM. That might have seemed the ultimate in career insanity a decade earlier, but IBM, now buried in antitrust lawsuits, was intentionally staying out of new industries and leaving alone new competitors it once might have crushed. IBM 360 architect Gene Amdahl had shown it could be done back in 1970, when he started his own eponymous mainframe computer company.

Three years later, Shugart wanted to pursue the same path—that is, to build a complete computer system, including a central processor, memory, and printer—but unlike Amdahl, he wanted to target the small business market with a low-cost machine. The timing was perfect: IBM, HP, Wang, DEC, and Data General were all pursuing similar "mini" computers and workstations for both business and scientific applications. Shugart had some clever ideas for how to compete in this market, but as he soon learned, he didn't have near enough capital to compete with these giants. And by 1977, he was out of money, without a finished product to take to market. He would later say that Al's Law of Business Number One was: "Cash is more important than your mother."[7]

Needless to say, this situation resulted in a confrontation between the company's founder and his investors. Shugart wanted to keep going; the venture capitalists wanted out. As is always the case in such fights, the money won. Shugart always claimed he walked out in frustration; the investors always claimed he was fired. He would later say, "Actually, I don't know if I got fired or if I quit. A friend told me later that, for a person in my position, the difference between firing and quitting is about five microseconds."[8]

The result was the same, and Al Shugart always said that the most painful experience of his life was having to drive every day past the company headquarters bearing his name knowing he was barred from ever again being allowed inside.

In short order, the investors sold the company to Xerox, which changed the subsidiary's name to Shugart Corporation. Xerox was notorious in those days for never failing to miss a good opportunity; just up the road in Palo Alto, at its research center, Xerox was about to invent not only the personal computer but also the mouse and the windows-type operating system . . . only to fail to follow-up on any of them. But with Shugart Corporation it actually spotted a prize and pursued it.

A FLEXIBLE SOLUTION

About the time Shugart Associates was being founded, IBM had embarked on a new program to come up with a cheaper, removable version of its now hugely successful Winchester disk drives. The company shrewdly turned to the other major memory medium—magnetic tape—for inspiration. With the disk-drive paradigm before them, the solution to the problem of sequence with tape now seemed obvious: Just cut the coated film (now Mylar) into a disk instead of ribbon. Now, if you put this disk into a more rigid holder to keep it from flexing, and rotated it like a 45 rpm record, you'd have a low-cost memory medium that could be read with a Winchester-type head, it would be light and removable, and it would cost just a few bucks.

IBM introduced the first 8-inch-diameter version of this flexible disk—it was soon nicknamed "floppy" for obvious reasons—and accompanying drive in 1970 and it was an immediate hit. The disks could be filled with data and removed and filed like folders—at a fraction of the cost of a hard-disk drive. Many small-business customers eventually stuck with the floppies and eschewed hard disks altogether.

By the time Al Shugart and his team were designing the Shugart computer, 8-inch floppies were ubiquitous in the computing world. But they were also showing their limitations. In particular, the very first microcomputers—the immediate precursors of personal computers—were beginning to appear for scientific and electronic-design applications. And it was hard to be "micro" when the peripheral floppy disk drive was the size of a large telephone book.

It was to meet this as-yet-unmet demand for a smaller floppy drive that in 1976 Shugart president Don Massaro and sales director Jim Adkinsson met with a major client to ascertain what he wanted in the next generation of floppies. In particular, they asked, what size do you want the disk to be? Because this was a sales call, and the three were meeting in a bar, the client pointed at a cocktail napkin and said, "That big." Massaro and Adkinsson took the napkin, measured it, and, finding a pair of scissors, cut out a matching square of cardboard on the way back to the office. They specifically tested it to make sure it was slightly too large for a shirt pocket—they were apparently concerned that the disks would get bent if carried around that way—and then presented this new 5.25-inch size to Al Shugart . . . who approved it immediately.

Within a year, Al Shugart was gone, but the new stripped-down, Xerox-owned Shugart forged ahead as the first 5.25-inch 360KB flexible-disk-drive company—and made a fortune. Existing computer users loved the new smaller size because it made it easy to carry programs from one machine to another or to quickly download work from memory for filing. But even more important, the personal computer revolution was now under way. Apple introduced the landmark Apple II in 1977—and within months had scores of competitors. For this first wave of personal computers, hard drives were out of the question: Not only were they as big as the computers themselves, minus the displays, but they typically also cost as much as the rest of the computer—all for about ten megabytes of memory storage. In the new 5.25-inch floppies, the PC makers found the perfect storage medium.

By 1978, Shugart Corporation had more than ten competitors, all building comparable drives. The race was now on to see how much data could be stuffed onto each disk; and that would be a function of the density of surface on the individual disk and the speed of the read-write head.

ODD MAN OUT

But the most influential figure in the disk-drive industry was out of the game. Unemployed, other than a few pickup consulting jobs, Al Shugart moved over the hill to Santa Cruz and the life of a prosperous beach bum:

"I bought a house on a cliff overlooking the ocean—wonderful place, pool and everything."[9]

With some partners, Shugart bought a bar in Santa Cruz and ended up spending part of his time slinging drinks or cleaning up at closing time.

"I had a good time. I bought a fishing boat and was fishing for salmon and albacore, and selling it. . . . My day started overlooking the ocean, hearing the water and so forth. I didn't have to be at work at eight o'clock in the morning, so I therefore could miss the traffic. I would go at ten or I could go at five. . . ."[10]

Al fished commercially out of Santa Cruz, and in time moved up to San Francisco Bay, where he would often deliver his daily catch to Fisherman's Wharf. The tourists, seeing the stocky man with the shock of graying hair hauling a load of fish over his shoulder, had no idea that they were looking at the man who already changed their work and would soon transform their lives.

"I always thought that I enjoyed life more than everybody else. So it doesn't bother me if somebody drives by in a Mercedes and I'm in an old fishing boat. I'm sure that I was enjoying life more in my old fishing boat. . . . I never felt sorry for myself. I think about the bar and the fishing boat sometimes."[11]

It all sounded good, and the sunrises over the Bay Bridge were beautiful . . . but for a born entrepreneur like Al Shugart, it was privately excruciating not to be back in the game, especially as he began to elaborate a vision of where computer memory needed to go next. Finally, in 1979, he teamed up with a group of industry veterans, including his old partner at Shugart Associates, Finis Conner, and founded a new disk-drive company. They called it Shugart Technology, only to hear from Xerox that the new start-up could not be named after its own founder. So they changed it to Seagate Technology and set up shop in the Santa Cruz Mountains, halfway between Al's beloved Santa Cruz and his despised Silicon Valley.

Shugart and Conner shared a common vision of where they thought computer memory needed to go in the personal-computer era—and it was back to hard disks, with their immensely greater storage capacity. The trick, they realized, was to figure out how to put a Winchester hard-disk drive into the now-established shallow-form factor of the current 5.25-inch floppy drive.

In 1980, Seagate introduced the ST-506, the first hard disk to fit into the standard PC disk-drive bay. It held five megabytes—ten times that of the standard floppy of the era—and was soon followed by a version that held ten megabytes. When IBM chose the Seagate drive for its IBM PC XT, the first personal computer from the company to use a hard disk, Seagate's fortunes were made. By 1993, the company shipped its 50 *millionth* drive; by 2008, it was one *billion* drives—and the 56,000 employee company had annual revenues of more than $10 billion.

But Seagate wasn't the only competitor chasing the fortunes of supplying hard-disk memory for the hottest consumer electronics product of the age. By the time this new disk-memory race was raging, Silicon Valley—and more important, Silicon Valley's venture capital industry—had matured into the most efficient incubator of new entrepreneurial start-ups the world had ever seen. A smart team with a good product idea could almost always find not just the capital they needed but also the personnel, the manufacturing, and the marketing they needed to ramp up fast.

It all came together at the beginning of the 1980s, setting off the biggest new company land rush high-tech had yet seen. Within twenty-four months after Seagate's announcement, an estimated 250 new 5.25-inch hard-disk-drive companies had been founded—all of them pursuing a dominant share of the market. Of course, that was impossible, and by 1997, an estimated 210 of those companies were already out of the business and most of them shuttered for good.[12] By the new century, the number of competitors was less than a dozen, with Seagate still standing as the world's largest independent disk-drive company.

DATO

By then, though, Al Shugart was gone again. In July 1998, he officially resigned all of his positions at Seagate—"I was fired," he said a few years later. "The board told me it was time for change. That was the only reason I was given."[13]

The only man to ever be fired from two billion-dollar companies that at some point bore his name, he now founded Al Shugart International, a boutique angel-investment/executive-consulting firm of less than a dozen employees that was largely a platform for Shugart to pursue anything that interested him. In a valley of characters, he was one of the most famous, and this last phase of his career allowed him to indulge his opinions and eccentricities without worrying about disapproving boards and unhappy shareholders.

"I have always been an independent cuss. That's part of being an entrepreneur. The only two companies I've ever been fired from were the two companies I started."[14]

Shugart took to wearing Hawaiian shirts, as did his staff—mostly pretty women. He reveled in being named a *dato*, a Malaysian honorific, and the respect it earned him in that crounty. A natural libertarian ("I object to politics generally"), he nevertheless ran his dog, Ernest, for Congress from the Monterey area in 1996 in order to shake up local voters from what he thought was widespread apathy. It drew national attention. So did Shugart's more serious initiative to officially add "None of the Above" to all California state ballots. He was also a pioneering supporter of simplified tax forms and campaign finance reform . . . all in an effort, he said, "to get more people to get more active in politics."

"I think I'm doing some good. If I didn't think I was doing some good, then I wouldn't like it. If the politicians don't like it, then I know I'm on the right track."[15]

Looking back in 2001, he mused, "I really enjoyed success, and not just my success. I like to do things well. Doing things well in the disk-drive business was very challenging, but we did things well. But I like other peoples' success too, and so when I see all these kids starting companies and becoming billionaires, I'm happy for them. That's success."[16]

Al Shugart, the most indelible figure of the computer disk-memory industry, died in December 2006 of complications from heart surgery. One of his last public images was a Christmas card showing the *dato* himself, grinning and surrounded by his staff, all wearing Hawaiian shirts.

The disk-drive industry continued, of course, and to even greater glory, but henceforth, robbed of its only celebrity, it would be all but anonymous.

The drives themselves grew ever faster, of greater capacity, and cheaper. In the race to keep up with the tireless demands of the personal-computer industry for more performance, some companies (such as Maxtor) tried the old trick, dating back to RAMAC, of stacking multiple disks inside a single player. In just eleven years, from 1980 to 1991, disk-memory technology advanced at a staggering pace: Al Shugart's Seagate ST-506, with its 5.25-inch disks and five megabytes of memory, had cost $1,500; less than a dozen years later, multiple companies were building 2.5-inch disk drives, containing one hundred megabytes of memory for half that price. A year after that, Hewlett-Packard raised the ante with a 1.3-inch disk drive—the size of a quarter—in a case not much bigger than a large postage stamp.

These breakthroughs had two important effects. First, it effectively killed the floppy disk as a standard memory format. Floppies, struggling to keep up all through the 1980s, introduced a 3.5-inch version in a more rigid plastic case that ultimately reached a 1.44-megabyte capacity. But though there were later attempts to create a higher-capacity version, the game was up by the mid-1990s, when Apple Computer, which had been the last holdout against an internal hard-disk drive, finally made the move. The second important effect of the newer, smaller, and higher-capacity hard disks was that they (along with flat-panel displays) made possible the revolution in laptop computing, smart phones, and other consumer products.

FREEDOM COMES FREE

Memory, which during the era of mainframes and magnetic core had been one of the most expensive and largest parts of the computer, had now become one of the cheapest and smallest. RAMAC's memory had cost users $150 per megabyte to *rent* per month; by 2000 that had been reduced to a *purchase* price of just $0.02. A decade later, it had fallen—on a two-terabyte, 3.5-inch, five-platter disk drive—to less than *one-thousandth* of a cent per megabyte. In 2011 it was possible to purchase a Seagate internal, 2.5-inch hard drive with a capacity of 750 gigabytes, for less than $100—a device so small it could fit easily into a handheld game player.

Meanwhile, it was slowly dawning on big Internet service companies, such as Google, that their giant server facilities around the world contained trillions of bytes of disk-memory storage that were essentially being unused most of the time. So they began to devise new services for this amorphous *cloud* of unused storage that their users could access essentially without cost. The first, and most famous, of these cloud applications were GoogleMail—Gmail—in 2004, and GoogleMaps a year later. Soon, new Internet cloud services were available from numerous companies, ranging from low-cost to more secure, but more expensive, versions.

This staggering drop in prices made possible by disk storage saw artificial memory become essentially free for the first time since the Renaissance memory artists—and this time, everybody could take advantage of the opportunity. That in turn created a paradigm shift in the relationship between human beings and stored knowledge that has only begun.

For one thing, "free" memory only accelerated the complexity arms race in software and applications that had been going on almost from the introduction of the first commercial computers. Even as computers had grown smaller and more personal, their growing processing power and memory size enabled them to add ever more performance—computation, bookkeeping, networking, word processing, spreadsheets, desktop publishing, games, personal communications, multiplayer games, communications, streaming video, 3-D graphics, and on and on—that only whetted the desire of consumers for even more.

PC owners in 1980 dreamed of owning a five-megabyte hard disk but wondered if they'd ever need that much storage; their children took the terabyte disks—capable of holding every written word in the world in Imperial Rome—that came with their laptops and worried that they might run out of memory for all of their games, videos, and photos. Free memory meant free rein for one's imagination for creating computer experiences.

At the other end of the scale, all of this free memory, combined with ever-faster processing speeds, liberated scientific researchers to imagine and then tackle tasks that were once almost beyond human imagination, such as modeling every air molecule in a storm or the neutrons in an atomic explosion, creating virtual realities that were indistinguishable to the human eye from the natural world, practicing medicine on fully functional "virtual" patients, mapping the entire human genome—or modeling the operation of memory in the human brain.

MISSING LINKS

But it was in between these two extremes of Big Science and small consumers within the everyday operation of companies, agencies, and universities, that the biggest and most important effect of free memory took place.

Almost from the moment there were two computers in the world, their operators have wanted them to talk to each other. There are, after all, a number of reasons for having computers "network": It cuts out the expensive and time-wasting middlemen of card and paper printers and readers; it allows computers to "talk" at the rocket speeds at which they operate; and it makes possible the sharing of tasks to cut down overall operating time. All of these advantages were best captured by Robert Metcalfe—himself the legendary coinventor (at Xerox in 1973) of the landmark networking protocol Ethernet—when he noted that the value of a computer network seems to increase at the square of the number of connections on that network. That is, the bigger the network, the very much bigger its usefulness.

It was an implicit understanding of this principle that had led researchers as early as the late 1930s to experiment with remote accessing of computers. The pioneer of this field was George Stibitz, a Bell Labs researcher who had already played a key role in applying Boolean algebra to computer circuits. On September 11, 1940, at a meeting of the American Mathematical Society at Dartmouth College, Stibitz used a teletype machine to direct some computational work on his self-made Complex Number Calculator computer back at his office in New York.

By the late 1950s, the U.S. military was using primitive networking to share data from its many radar-control systems, while a pair of mainframe computers, owned by American Airlines, were linked together to create SABRE, the forerunner of the modern airline reservation system.

This was followed by a series of breakthroughs in the mid-1960s that would make global networking possible at last. The first of these, in 1964, came out of Dartmouth, where a team of researchers created "time-sharing"—the ability of multiple remote users (typically armed with an acoustic coupler modem for their telephone and a teletype machine) to take turns accessing a remote computer. Time-sharing would prove to be the inspirational first experience that many of the pioneers of the personal-computer industry—notably Steve Wozniak—would have with "home" computing. It was in a quest to duplicate this childhood experience that many fabricated their first computers.

At about the same time, Joseph Carl Robnett "Lick" Licklider, a brilliant multidisciplinary scientist, presented a paper, which he entitled

"The Intergalactic Computer Network," to the employees of the U.S. Department of Defense Advanced Research Projects Agency (ARPA), which he would soon join as a director. The title of the paper was meant to be a joke, but its core message wasn't. And among the young scientists it inspired was program director Lawrence Roberts, who turned his team to the task of creating such a global network. Meanwhile, working in parallel on a way to cluster data into "packets," transmit them over the shortest available network pathways, and reassemble them at the target location, were Paul Baran at the RAND Corporation and Donald Davies at the UK's National Physical Laboratory.

It was Davies who coined the term "packet switching." And when, in 1969—the *annum mirabilis* of the digital age—ARPA (now DARPA) set out to tie together the computers at government agencies, research laboratories, and universities into a common network called Arpanet, it was Licklider's vision, Roberts's networking architecture, and Baran's and Davies's packet switching that made it all work. Their inventions would enable that network to grow in the 1980s with the help of a final critical invention, by Robert Kahn and Vinton Cerf, of the Internet Protocol Suite—TC/IP—into the *Internet.*

It would also be Baran's and Davies's packet switching that would make possible the global cellular telephony industry. Baran would go on to become a Silicon Valley legend, founding four companies that were each valued at more than $1 billion. He was working on his newest company the day he died of cancer, at age eighty-four, in 2011.*

DISK TO DISK

For the billions who today use the Internet on a daily basis, the Net seemed to spring fully formed on their computers in the mid-1990s. But the forgotten decade preceding that global rollout was vitally important. And it depended heavily on the ongoing race to build ever more powerful disk drives.

In operation, the Internet requires a hierarchy of computers—both small and large—to function. The small computers, mostly PCs and smart

---

* The author was a cofounder with Baran of this last company.

devices, act as the access points to the Net; the large computers sit at the crossroads of the data flowing around the network and manage the traffic. That's the simple version. A more accurate description is that the disk memory at the individual nodes (managed by special software in the PCs) communicates with big disk drives (enterprise hard disks), managed by specialized computers (servers) organized by the scores in big warehouses (server farms) and communicate with other specialized computers designed to manage the flow of data, not process it (routers).

Thus, another way of looking at the rise of the Internet in the 1980s is that it is the story of getting sufficient processing and memory power into home and office computers via small, inexpensive disk drives; building full-size disk drives powerful enough (high-capacity, very fast access speeds, and 24/7 reliability) to manage huge data flows; inventing the new routers and other hardware—the best-known manufacturer being Cisco—needed to manage this infrastructure; and developing the standards, software, and applications required to make all of this work smoothly and, for end users, intuitively.

For the end-user experience, there were several key players who built upon the work of their predecessors. The first was Tim Berners-Lee, a scientist at the European CERN nuclear research center in Geneva, Switzerland, who, in 1991, first proposed the simplified Internet addressing architecture that became the World Wide Web—and made the Internet at last accessible to consumers. A year later, a team at the National Center for Supercomputing Applications at the University of Illinois at Urbana–Champaign embarked on a project to create a graphical "browser" to simplify access to the growing number of websites on the Internet. The result, Mosaic, was introduced in 1993. Almost immediately, a team of code writers from the Mosaic project, led by Marc Andreessen, teamed with workstation tycoon Jim Clark to start Netscape.

Netscape's Navigator browser proved so popular that the company became a business superstar like Apple, Intel, and Ampex before it—which was enough to capture the attention of the biggest software company in the personal computing world: Microsoft. Bill Gates and his team in Everett, Washington, who by now utterly dominated PC operating systems, had been caught flatfooted by the Web and Netscape Navigator. So they set out, by any means necessary (including bundling its own new browser into the Microsoft Windows operating system), to crush Netscape.

Microsoft succeeded, at the cost of a federal investigation. But it would fail to do the same with the next great Web application. This was the "search engine," which appeared in many forms in the 1990s in response to the need to manage the growing list of thousands of new websites cropping up every month.

Nearly all of these early search engines had a fatal flaw, however. Many prioritized their searches weighted by their advertisers. Others, such as the otherwise hugely successful Yahoo!, gave priority to its own select list of sites.

In the end, it was two Stanford students, Sergey Brin and Larry Page, who came up with a search engine that was organized only by the congruence of sites to the search question, and by number of visitors. This proved to be the magic recipe for search, and Brin and Page's company, Google, was founded in 1998. Their single most important executive decision was to hire Eric Schmidt, one of Silicon Valley's most brilliant technologists, to run the company. Schmidt, who had been beaten in the marketplace twice by Microsoft (at Sun Microsystems and Novell), came up with a strategy to hold off Gates and company—and managed to pull it off. A decade after its founding, Google still owned more than a 90 percent share of the world's Internet searches. Google became the defining firm of the era, one of the most valuable in U.S. industry, and the template for the dot-com boom that began at the end of the 1990s and firmly established the Web as an inextricable part of the lives of most people on the planet.

THE LAST TRACK

Jon Rubinstein, Apple Computer's chief of hardware engineering, faced what seemed an impossible challenge. Apple's cofounder, the brilliant and mercurial Steve Jobs, had returned to the company after a twelve-year hiatus just a few years before and had quickly revitalized Apple with a series of astonishing new computer designs. But now, in 1996, Jobs wanted to turn the company's attention toward other emerging opportunities in the consumer hardware business.

The public often thinks of Steve Jobs as a genius inventor like the young Tom Edison—something neither Jobs nor Apple did anything to correct. He never was one, in fact, and usually depended upon others of greater technical facility—Steve Wozniak, Jef Raskin, and others—to do the creating. Instead, Jobs was more like the older Edison: an impresario of invention—perhaps the greatest ever—setting out a vague idea of what a new product should be; creating an environment that supported risk-taking, attention to style, and the user's experience; and then using his own reputation and charisma to give Apple unmatched marketing power.

One of the secondary effects of the disk-memory race and the rise of the Internet in the 1990s is that it made it possible for the first time for users to easily swap very large chunks of memory—games, images, and, most important for college students, music files. Swapping music files violated copyright laws, but students, driven by the technical imperative that "any good new technology will find its users," flouted the law by the millions, especially when new websites—most notably Napster—emerged to simplify the process.

Soon the music industry was suing Napster while the FBI was arresting some of the more egregious music file pirates. But as with Prohibition seventy years before, the craze only went underground . . . and grew. Steve Jobs watched this trend and, while other big companies kept their distance, he saw a gigantic opportunity. Jobs was no stranger to illegal activities—he and Steve Wozniak had begun their tech careers as sellers of illegal telephone hacking equipment—so he had a unique perspective on how the piracy mess would resolve itself, and he planned to put Apple right in the middle of that solution.

Jobs knew that to do so would require a two-part strategic play. On the one hand, he had to co-opt the increasingly paranoid and litigious music industry, which was watching its once-hugely profitable business being crippled by a new technological paradigm, the MP3 music file, and an entire generation of young bootleggers circumventing the rules of copyright and the marketplace. His solution would be to use his leverage as chairman of the motion picture company Pixar and as the most famous figure of the consumer digital revolution to propose a compromise: the creation of a legal online shop for downloadable music files, which he would call "iTunes," that would charge a fee for songs that, while low by music-industry standards, would also be cheap enough to convince millions of young people to abandon their criminality and turn to a legitimate source.

But that was only half of the strategy. Even as Apple positioned itself as the key content supplier for digital music files, it also wanted the hardware business as well—a brand-new industry whose clunky, oversized products were perfect targets for the Apple style.

And that's where Jon Rubinstein entered the story. As Jobs explained it, he wanted a device small enough to fit into a shirt or jeans pocket, with an elegant touch control and small but crisp screen, a headphone jack, a nonremovable battery that could be recharged on a tiny dock, and a standard Apple FireWire interface to enable the device to download from an Apple computer (and also recharge from it). All of that would be tough, but doable, thought Rubinstein. But then the clincher: Jobs wanted this device to cost only a couple hundred dollars, while still able to put—as Jobs would eventually say—"A thousand songs in your pocket."[17]

Rubinstein gulped on that last bit. He had followed the memory industry long enough to know that there wasn't a single micro hard-disk drive in the world capable of fitting into a case that size while still having the gigabytes of memory needed to hold that much information. The only good news was that Jobs, who had been notorious in the past for backing the wrong type of memory (for example, the experimental laser-based "magneto-optical" memory drive in the NeXT computer), had left the choice up to Rubinstein.

So as he started the program, one of the first tasks Rubinstein set for himself was to find a manufacturer that would be willing to build a drive to these unique specs: a disk of less than two inches in diameter, in a drive of no more than two inches wide with proprietary connectors, capable of holding one gigabyte of memory. Rubinstein knew that this was asking the almost impossible, but he assumed the Apple name would at least spark some interest among the two dozen disk-drive companies in the world.

Boy, was he mistaken. He was dismissed, laughed at, met with stunned silence, and even on one occasion hung up on by a major manufacturer who thought it was a prank call. After having no luck with the first tier of manufacturers, a desperate Rubinstein started calling the also-rans. In the end, the only company to show interest was the big, diversified Toshiba of Japan. Big as it was, Toshiba was the least likely of suppliers: It was only in the disk-drive business (mostly through a relationship with Fujitsu) in support of its personal computers and servers, and it had zero reputation in disk memory for innovation.

Rubinstein realized, however, that Toshiba was the only game in town, and the Japanese giant got the contract. Happily for Rubinstein, Toshiba delivered the new little minidisks on time.

The resulting product, the result of Steve Jobs's vision and Jon Rubinstein's pragmatism, was, of course, the Apple iPod—the first great product of the new age of consumer electronics, setting the stage for the iPhone and iPad to follow. It was the start of perhaps the greatest run of landmark new consumer products since Edison himself. Introduced in late October 2001, the iPod got off to an almost invisible start thanks to slow delivery to retailers, an economic recession, and the distraction of the 9/11 terrorist attacks in New York City. But by the end of the decade, the iPod was a phenomenon of historic proportions: As of October 2011, *320 million* iPods in various models and configurations had been sold by Apple Computer.

But what had been seen as the ultimate triumph, the zenith, of the hard-disk drive would prove in time to be the beginning of the end of its era. The early generations of the iPod "classic" design would continue to use the Toshiba 1.8-inch drive. And when Apple decided to downsize the device with the iPod "mini," it too would contain a hard disk—this one just an inch in diameter—from Hitachi and Seagate.

But that was it. In September 2005, Apple introduced the iPod Nano—a tiny MP3 player that was half the size of the original, yet still contained up to four gigabytes of memory. But there would never again be a disk drive in the iPod or any of its successors. Now the memory of choice would be "flash" memory chips. After fifty years of chasing magnetic memory, semiconductor memory had (for most applications) caught up at last.

IN THE CHIPS

The history of the semiconductor industry is the best known in tech, probably because it is the most venerable, the technology is fundamental to everything else in electronics, and most of all, because it contains the most remarkable characters.

But within that larger tale lies a number of other, less-well-known narratives, not least that of semiconductor memory.

A quick history: The semiconductor revolution, the defining technology driver of the twentieth century and beyond, began with a lecture in

1940, given at Bell Laboratories in New Jersey. There, the speaker, researcher Russell S. Ohl, began by showing a small slab of silicon with a wire attached at each end. Ohl then shone a flashlight onto the middle of the silicon . . . and to the amazement of the assembled scientists, electrical current suddenly passed through the glass, normally a natural insulator. A circuit had opened, Ohl explained, because the silicon wasn't pure, but rather "doped" with impurities like boron and phosphorous from the third and fifth columns of the periodic table.

Ohl went on to explain that when the energy from the flashlight beam had hit the center of the slab, these dopants had given off electrons in such a way that the silicon had become a conductor—a kind of "gate" that closed again when the light was switched off. Because of these attributes, Ohl called this doped silicon a "semiconductor."

Two of the scientists in the audience, John Bardeen and Walter Brattain, walked out of Ohl's demonstration convinced that they'd seen a possible answer to the biggest practical electronics challenge of the day: creating a replacement to the vacuum tube, which was becoming too delicate, too slow, too hot, and too energy hungry for the growing number of tasks and environments in which it was being used. Bardeen and Brattain agreed that if they could create a functional, solid-state on-off switch using this new semiconductor technology, they would revolutionize electronics.

But before they could get started, World War II broke out and both men were assigned to other, more immediate concerns. Ironically, the war demonstrated more than ever—on the battlefield, in warships, in airplanes, in jungles, in deserts, and in snow—that the world desperately needed a replacement for the vacuum tube.

With the war's end, Bardeen and Brattain finally got back to their project, and over the next two years they labored to create a workable semiconductor circuit. Ultimately, facing some recalcitrant problems with the physics of the device, they turned to one of their compatriots for help. That scientist, William Shockley, was considered not only the most brilliant scientific mind at Bell Labs, but some said one of the greatest since Newton. It was with Shockley's help that Bardeen and Brattain finally built a working circuit on December 23, 1947. It looked like a tiny arrowhead of quartz embedded into a slightly larger sliver of germanium, with both components trailing wires. Electricity flowing into the quartz arrowhead acted as a valve on electricity passing through the germanium sliver, turning it on and off in the ones and zeros of computing's Boolean algebra.[18]

This was the *transistor,* often hailed as the most important invention of the twentieth century, and when it began to appear in commercial applications in the early 1950s, the little germanium junction was hidden under a metal cap ("can") atop a tripod of "leads" that were extensions of those three controlling wires. Bardeen, Brattain, and Shockley were rightly awarded the Nobel Prize.

In the story of memory, the invention of the transistor can be compared to that of the book, or even printing, in terms of its influence. And what made it astounding was that it was so simple. As Gordon Moore, one the most famous figures in the history of electronics, would point out, part of the miracle of the semiconductor device was that it was literally so elemental. Made of one of the most common substances on the planet, silicon sand, it was forged out of fire, rusted with oxygen, and purified with water—hearkening back to the pre-Socratic Greek philosophers and their belief that the universe was composed of fire, earth, water, and air.[19]

That is to say, the heart of the transistor was as tough and enduring as the rock it was made from, which meant that, unlike De Forest's tube, it could endure great heat and cold, the pressure at the bottom of the ocean, and the vacuum of outer space—even, sometimes, the radiation of an atomic bomb. Left alone, it was almost immortal, vulnerable only after centuries to the effect of cosmic rays. Almost as important, it required little power to operate and gave off comparatively little heat.

The transistor—replacing tubes in everything from mainframe computers to test and measurement instruments to portable radios—transformed the world of electronics by making possible smaller, more durable, and more efficient devices than ever before. It also set off a gold rush of new and old companies chasing the potentially unlimited wealth to be made as a transistor manufacturer.

One of these new transistor companies was founded in 1956 by Bill Shockley himself. He had left Bell Labs and come home to Palo Alto to be close to his ailing mother and to start his own company, Shockley Semiconductor Laboratory. There Shockley planned to improve upon the Motorola Company's newly discovered use of silicon as a replacement for the more expensive germanium.

Such was Shockley's reputation that, when he put out word that he was looking for top talent for his new company, he was deluged with applications. In the end, he selected the eight most talented young physicists, chemists, and electronics engineers from around the United States to join him and build the world's finest transistors.

But while Shockley may have been a great scientist, he was a terrible boss, and he only grew crazier (with his racist IQ theories and "genius sperm bank") as the years went on. It wasn't long before his eight young scientists had had enough of his paranoia and belittlement and conspired to resign en masse and start their own company.

Eventually, in a search process that would create the modern venture capital industry, the "Traitorous Eight" (as Shockley called them) would find an investor in the defense contractor Fairchild Camera and Instrument, after the leader of the Eight, Robert Noyce, made an impressive and impassioned speech to Sherman Fairchild about the potential for silicon chips. It would be Noyce who would not only lead the new Fairchild Semiconductor but devise the technology that would soon make the little Mountain View, California, operation into the most important company of the postwar world.[20]

THE PLANAR TRUTH

By then, 1957, change was in the air in the semiconductor world. Five years earlier, British scientist G. W. A. Dummar had predicted:

> *It seems now possible to envisage electronic equipment in a solid block with no connecting wires. The block may consist of layers of insulating, conducting, rectifying and amplifying materials, the electrical junctions connected directly by cutting out areas of the various layers.*[21]

Almost from the day of Fairchild's founding, the Traitorous Eight—especially Bob Noyce—were already pondering how to make Dummar's vision real. But Fairchild Semiconductor wasn't the only enterprise pursuing advanced transistor technology. Another was Texas Instruments. There in the sweltering summer of 1958, when veteran employees were allowed to leave the office in the heat, a new employee named Jack Kilby was required to stay. Bored, he decided to write down in his journal some notes on his solution to Dummar's ideas. By the end of that summer, he built a prototype for this germanium circuit and earned a patent for his design.[22]

At almost the same time, Noyce and his team at Fairchild were pursuing their own vision of this circuit, this time in silicon and based on a very different Noyce design. Like Kilby, Noyce understood that if this circuit could be made as a flat "sandwich" of silicon or germanium and metal conductors, it might ultimately be possible to put more than one such transistor on a single chip and link them together—that is, into an *integrated circuit*. Noyce divided his small staff into two teams, one under Jean Hoerni, the other under Gordon Moore, and set them to work finding a viable way to fabricate this design.

It was Hoerni who came up with the solution, which he called the "planar process," and it would define semiconductor fabrication for the next half-century. The breakthrough of the planar process was not just that it achieved the flat structure needed for the integrated circuit but that it did so with a manufacturing process that was most akin to printing. A thin wafer of silicon was coated with a photoreactive material much like that used by Talbot in photography more than a century before, and then exposed by ultraviolet light passing through a stencil-like mask containing the image of one layer of circuitry. The unexposed photoresistor was then washed away with acid, and the remaining image was cooked into place in an oven. This process was then repeated with the next layer of circuitry—and so on, up to twenty or more such layers. Then a layer of metal conductor was plated onto the surface of these many layers, reaching down through holes to the lower layers, to create the equivalent of interconnecting wires.[23]

What made the planar process so important was this same photolithography technique could be used to put not just one but multiple interconnected circuits on the surface of a silicon wafer.

The first Fairchild planar transistor—called a "mesa transistor"—looked like a tiny bull's-eye. IBM bought 150 of them, as much to study as anything else. But within two years, Fairchild had managed to stuff four transistors on the surface of a chip . . . and the number was soon doubling at a breakneck pace.

History would recognize Noyce and Kilby (the latter winning a Nobel Prize after Noyce's early death) as the coinventors of the integrated circuit—the transistor's descendant and rival for the title of Invention of the Century. But it was Hoerni's breakthrough with the planar process that made the subsequent "computer chip" revolution possible.

By the early 1960s, Fairchild, armed with its IC technology, was the hottest young company tech had seen since Ampex a decade before. It was also perhaps the greatest collection of young entrepreneurial talent ever assembled. They were brilliant, talented, young, and wild—and the hard-drinking, skirt-chasing, rule-breaking Fairchild soon gained a reputation, that still stands, as the wildest company in Silicon Valley history. An eternal "what-if" in high tech is to ask what would have happened if that original Fairchild crew had managed to stay together, given that its employees would go on to create a trillion-dollar semiconductor industry as well as play key roles in other industries of almost equal size, including computer games, cellular telephones, displays, and personal computers.

But Fairchild was just too volatile to remain intact for long . . . and when, in 1967, the parent company refused to grant stock options and allow its California employees to share in the riches they'd created—and wasted much of that division's profits on failed ventures—Fairchild Semiconductor shattered, scattering talent all over the Valley that eventually coalesced into an estimated one hundred new chip companies, including Intel, National Semiconductor, Advanced Micro Devices (AMD), and Zilog. With this explosion of new chip companies, the modern Silicon Valley was born. Visiting reporter Don Hoeffler, noticing all of these new chip companies, gave the place its name.

CHANGE AS LAW

By the time of Fairchild's great hemorrhage of talent, the semiconductor industry had already begun to divide into separate market sectors, largely congruent with the separation of operations within computer architecture. Thus, one part of the semiconductor industry pursued the *logic* chips used in computer processors (this was Fairchild's specialty); *memory* chips, used to provide on-board storage for regularly used information that didn't go out to the disk drive; *input-output* (I/O) chips, often containing both digital and analog circuits to manage the flow of data in from terminals and other sources and out to printers and other networked computers; and *linear* or analog chips (such as diodes and resistors) that handled the flow of electricity around those other chips.

Of these four types, only memory chips seemed to progress at a steady rate. They were more monolithic in their design, lending themselves to greater miniaturization, and they didn't require the complex fabrication of I/O chips or the burst of individual design genius typically found in the linear world. And they were more universal in application than logic chips, whose primary market at the time was mainframes and minicomputers.

So systematic was the onward march of memory chips that it caught the attention of Gordon Moore at Fairchild. In 1965, having been asked to write an article for *Electronics* magazine, Moore sat down with a piece of graph paper and began to plot the performance of chips versus the date of their introduction. He chose memory chips because of their success, and quickly switched to logarithmic paper when he realized just how fast the progress had been. Moore had only a few data points—the integrated circuit was only seven years old at this point, and the most powerful memory chips at the time only held about 64 transistors—but a trend was already clear. Even then, Moore was stunned to see that the points on his graph were arrayed in a straight line. Integrated circuits, it seemed, were doubling in performance (capacity, miniaturization, price) every eighteen months. If this trend continued—and there was no reason it wouldn't, Moore wrote—this endless doubling (like the grains of rice on the chessboard in the Chinese tale of the clever man requesting payment from the emperor—2, 4, 8 grains, and so on) would result in unbelievable gains in the years ahead. Moore predicted that by 1975 a single memory chip might hold 64,000 transistors.[24]

History proved Moore's prediction to be uncannily accurate. By then, this doubling of chip performance every eighteen months (it would later slow to twenty-four months) was being called "Moore's Law." In fact, it wasn't really a scientific law—like, say, Metcalfe's description of the power of growing networks—but rather a kind of social contract between chip makers and their customers . . . and eventually with humanity . . . to maintain this doubling as long as possible with every bit of investment, management focus, and creativity it could bring to bear. The world, in turn, tacitly agreed to buy each of these succeeding generations of chips at a premium price and use them to drive newer and more powerful generations of consumer, industrial, and military products.

This unlikely relationship, between the chip industry and everyone else, has proven to be—in terms of advancing human wealth, health, and innovation—one of the most fruitful in history. Gordon Moore had hoped that his law might last a decade. Now, a half-century later, with semiconductor companies still struggling to maintain its momentum, it can be said that the global economy has now become the very embodiment of Moore's Law:

> *Today . . . it is increasingly apparent that Moore's Law is the defining measure of the modern world. Every other predictive tool for understanding life in the developed world since WWII—demographics, productivity tables, literacy rates, econometrics, the cycles of history, Marxist analysis, and on and on—have failed to predict the trajectory of society over the decades . . . except Moore's Law.*
>
> *Alone, this oddly narrow and technical dictum—that the processing speed, miniaturization, size and cost savings of integrated circuit chips will, together, double every couple years—has done a better job than any other in determining the pace of daily life, the ups and downs of the economy, the pace of innovation and the creation of new companies, fads and lifestyles. It has been said many times that, beneath everything, Moore's Law is ticking away as the metronome, the heartbeat, of the modern world.*[25]

Moore's Law did for the semiconductor industry something that had never happened before—it determined the pace of change for a generation ahead. And that in turn enabled entrepreneurs to develop new products and start new companies with the sure knowledge that the underlying technology and the not-yet-existent market would be waiting for them when they arrived. That proved to be true with the calculator and digital watch, the personal computer, the computer game, cellular telephony, the Internet, digital audio and video, medical devices, intelligent control of machines, virtual reality, and on and on. At this very moment, thousands of entrepreneurial teams around the world are devising business plans based upon the continued rule of Moore's Law.

THE INVENTION OF INVENTIONS

The explosion of Fairchild, and the resulting diaspora of semiconductor talent (the "Fairchildren") around the region not only created the modern

Silicon Valley but also established a field of ferocious companies whose competitiveness guaranteed that Moore's Law would get a roaring start. The most famous of these chip companies was Intel Corporation, founded by Noyce and Moore in 1968. This pair soon grew into a troika by the celebrated executive and scientist Andrew Grove.

Intel set out to use a new semiconductor technology (MOS) to become the world's leader in the fabrication of memory chips—and quickly reached that goal. But within a year, a major new opportunity appeared that the company, despite every effort, could not ignore. The electronic calculator boom was just peaking and was about to kill off all but the best-run competitors. One of these also-rans, the Japanese company Busicom, feared it would be one of the losers and decided to roll the dice on a radical new design. It approached Intel in October 1969 with the notion of putting multiple types of semiconductor circuits—logic, I/O, and memory—on a single chip. It had never before been tried, but like the IC itself a decade before, the idea was in the air.

Intel took the job, using its own scientist, Ted Hoff, to come up with the overall architecture of the chip (it would ultimately be four chips), which he based on the DEC VAX minicomputer, and Intel software expert Stan Mazor. To this team Intel added Masatoshi Shima, Busicom's top scientist and, from Fairchild, the world's most respected MOS designer, Federico Faggin.[26]

It was Faggin's arrival in April 1970 as the development team leader that put the project into high gear, and by the end of that year the Intel team had created a four-chip set capable of handling all of the calculator's operations, replacing three times as many traditional chips. Intel designated this set the Model 4004. It was the world's first working *microprocessor.* Today, with more than 20 billion in use around the world, providing intelligence to everything from phones and computers to rockets and robots, the microprocessor is the third candidate, after its ancestors the transistor and integrated circuit, for the Invention of the Century.

But it didn't start that way. The computer-on-a-chip approach was so radical that the 4004, and the later 8008, were initially met by customer skepticism. Companies around the world had converted from tubes to transistors to ICs pretty easily because they all did the same thing and they were all "discrete" (that is single-function, stand-alone) devices. But the microprocessor was a whole different way of seeing digital intelligence, and companies were wary of taking the risk. Intel itself considered abandoning the technology—not just because it wasn't taking off as planned but because from the beginning it had been a distraction from the company's core memory-chip business.

But then Faggin and his team created the Model 8008, a single-chip microprocessor . . . and suddenly the value of the microprocessor became clear both to established companies and new start-ups looking to leapfrog the competitor. The Model 8008 was followed by the Model 8080—the seminal device for all future microprocessors (the modern Intel and AMD chips are its direct descendants)—and when IBM picked a budget version of the subsequent 8086, the 8088, to put in its first IBM PC, the age of the microprocessor began.

Soon, Intel faced another dilemma. By the early 1980s, it had become the world's leading manufacturer of microprocessors. But it was also still the leading memory chip company—a hugely profitable business. The company was becoming schizophrenic, with both the processor and memory businesses vying for dominance. Meanwhile, other microprocessor businesses, from new companies like Zilog to established giants like Motorola, had jumped into the game and competition was ferocious. At the same time, the Japanese electronics giants had thrown their fortunes behind building memory chips and were now producing devices of such quality and low price that they were embarrassing the U.S. semiconductor industry.

It was becoming increasingly obvious that Intel had to pick one business to pursue. It was also obvious that while most of the company wanted to pursue microprocessors (technology leadership, better profit margins, defensible market), the two men at the top, Moore and Grove, wanted to stick with the more proven memory market. In the end, they caved (and were embarrassed ever after for being so stubborn), and Intel went into the microprocessor business alone . . . and by the late 1990s was, based upon its stock price, the most valuable company in the world.

MEMORY MOVES EAST

Intel's departure from memory essentially turned that industry over to the Japanese and their soon-to-be competitors in South Korea and Taiwan, and the chip companies in those countries scrambled to stake out their turf in the many different submarkets of the increasingly fractionated memory-chip world.

Memory chips now were available in two basic types: *volatile*, which meant that the chips had to be continuously powered, if only at a low level, to retain their memory contents; and *nonvolatile*, typically of lesser capacity, which retained their contents when turned off. Within these two categories there were also numerous memory types, such as SRAM and DRAM, PROM and EPROM. There were also experiments with other, more exotic technologies, such as magnetic "bubble" memory, but they proved impractical.

Volatile memory chips, invented first, ruled the technology world. The DRAM (dynamic random access memory), invented in 1966 at IBM, was long the gold standard on which Moore's Law was tracked, and during shortages, such as in the late 1970s, actually created an underground black market for companies desperate for those chips to power their products. And it was an accusation of price-fixing on DRAMs by Japanese semiconductor makers in the early 1980s that led to the Japanese–U.S. trade war.

The fundamental weakness of dynamic memory was that it required some kind of electrical source—typically a battery in consumer devices—to keep the chips from self-erasing as they shut off. As with most high tech, this acceptable compromise soon became unacceptable with widespread use. The race was on then to develop memory chips that would retain their contents when turned off.

The solution came from the other side of chip memory use—read-only memory, that small part of a computer's architecture that held permanently recorded programs to run the system and that would never be erased. Programmable read-only memory (PROM) chips had been invented in 1956 by Wen Tsing Chow of American Bosch-ARMA Corporation for the U.S. Air Force. PROMs worked by attaching a digital "fuse" to each transistor, locking down its position—open or closed—as the device was turned off. PROMs were expensive and difficult to use because of their permanence, so they were limited in their application to a very precise niche in the computing world.[27]

But that began to change in 1971, when Dov Frohman, an Israeli scientist (and later vice president) at Intel, invented the erasable PROM.

The EPROM was both nonvolatile *and* easy to program, erase, and reprogram. In other words, it began to close the gap between the two worlds of memory chips. Gordon Moore would claim that the EPROM was as important to the development of the personal computer as the microprocessor was . . . and its success was one reason why Moore and Andy Grove were so resistant to taking Intel out of the memory business.

There was one more step: the electrically erasable PROM, or EEPROM. The EEPROM was also invented at Intel, in 1978, by George Perlegos, but was perfected elsewhere when Perlegos and other scientists left Intel to form Seeq Inc. The critical advantage of the EEPROM was that unlike the EPROM, which had to be removed from the device to be reprogrammed, EEPROM could be erased and recoded in situ, i.e., while it was still in the device, by electrical signals.[28]

IN A FLASH

The next step from EEPROM was a short but hugely influential one. Technologically, the EEPROM was the definitive solution to the challenge of usable memory chips, but in practice it had some serious obstacles, especially in consumer electronics: It was expensive and it was slow. In 1980, Fujio Masuoka of Toshiba, responding to the growing need by his company for a new kind of EEPROM to use in its consumer devices, redesigned the EEPROM to sacrifice some of its performance for improved erase and reprogram speed. In particular, instead of working with individual bytes of memory on the chip (needed by computer companies), Masuoka designed his new device to erase and program in large blocks of data—resulting in markedly improved response times. One of his colleagues, seeing this new EEPROM in action, dubbed it *flash memory* because its speed reminded him of a camera flashbulb.[29]

By 2006, flash memory—now used instead of the slower and more fragile disk drives in digital cameras, smart phones, electronic tablets, and the ubiquitous memory stick/"thumb drive"—had grown into a $20 billion industry, or one-third of the world's entire memory-chip business. It had also made the Japanese and South Korean companies that built flash chips, such as Toshiba and Samsung, into major competitors on the world semiconductor scene once again.

In 2005, Toshiba, working with U.S. memory-card maker SanDisk, announced the first one-gigabyte flash chip. Later that year, Samsung announced its own chip with twice that capacity, proving that flash was exhibiting the characteristics of Moore's Law. A year later, Samsung introduced a four-gigabyte flash chip, its capacity the equivalent of the standard small laptop disk drive.

The news stunned the electronics world: Chip memory had always been the smaller and more expensive counterpart of disk memory. But now an important technological *and* cultural threshold had been crossed. Very few consumer applications of technology required more than a few score megabytes of storage. With flash chips now offering a thousand times that much memory storage, who needed disk memory anymore? Sure, a disk could hold a trillion bits of data, but it was also slower than a chip, and because it was an electromechanical device full of spinning and moving parts, it was also more likely to eventually break down.

Apple by now had long since abandoned disk memory in its iPods and iPhones for flash without any apparent loss of performance in the eyes of consumers. Other companies were following suit. And then the turning point: In June 2006, Samsung announced the first line of PCs that substituted flash memory for a hard-disk drive. Dell Computers announced a comparable line a year later. And while some computer makers covered their bets by offering hybrid systems that combined a disk drive with an attached flash memory cache, it was clear that the era of magnetic memory was coming to a close.

The age of solid-state artificial memory had begun.

10
The Persistence of Memory
Memory as Existence

In a 2005 speech, the CEO of Google, Eric Schmidt, for the first time offered an estimate of the size of the Internet in total bytes of memory.[1]

He put the total at 5 million trillion bytes—or, based on the average size of a computer byte—about 50,000,000,000,000,000,000,000 (50 sextillion) bits. All of this memory, representing a sizable portion of *all* human knowledge and memory, was stored virtually in more than 150 million websites, and physically in an estimated 75 million services located around the world. None of these numbers were accurate, other researchers added, and some might be off in either direction by a factor of five.

Of this total size of the Internet, Schmidt estimated that Google, by far the world's leading search-engine company, had managed after seven years to index 200 trillion bytes (terabytes), or just .004 percent of the Net. Most of the rest, he admitted, was essentially *terra incognita*—a vast region of unexplored data that might never be fully known. At the current rate, Google would need *300 years* to index the entire Internet—and that, Schmidt added, was only assuming the impossible: that the Internet wouldn't grow by a single byte in those three centuries. In fact, the Net probably grew by several trillion bytes just during the course of Schmidt's speech.

By 2010, these unimaginable contents of the Net were accessed by just short of 1 billion personal computers, nearly 700 million smartphones (the total for 2011), and several hundred million other devices large and small, in the hands of an estimated 2 billion users worldwide.

Some of these users, mostly from the developed world, arrived in cyberspace using powerful computers and handheld devices, linked via wireless networks or broadband cable, and stored hundreds of gigabytes of memory of their own.

Others, newer to the Web and often from developing nations, had reached the Internet any way they could: dial-up modems, cell phones rented from corner stands, desktop computers stationed in classrooms and local libraries, Internet cafés of the kind long gone from the West. But they'd made it at last, and whether they were selling goods on eBay or following bloggers covering events their censored national media wouldn't touch or taking online classes at distant universities they would never see, they were the first generation to have access to the world's accumulated memory. And because of that, they inhabited a unique new reality that none of their ancestors had ever experienced. For the first time, these billions (and 2 billion more are expected to join this global conversation within the next decade) had access to almost everything every human has ever known. And it was at their fingertips. And it was as good as free.

MOVING OUT

Memory is the guardian of all things. So wrote the author of *Rhetorica Ad Herennium*. That author, even if he was a master of human relations like Cicero, could have never imagined a world in which the hoi polloi, even the people he considered slaves, could have access to something almost indistinguishable from omniscience. Nor could Isidore of Seville, for all of his knowledge of men's souls. Nor Aristotle, for all of his vision of how the world works. Nor Giordano Bruno, whose memory theater, in the end, was an attempt to achieve this kind of universal knowledge. Not even Gordon Moore, as he sat with his graph paper and extrapolated out the future of technology and watched the curve go vertical, could have guessed the revolutionary shift in mankind's relationship to its own memory that would happen in his lifetime—and in which he would play a central role.

The old alchemists—Bruno, Paracelsus, Roger Bacon, even Isaac Newton—famously searched for the "Philosopher's Stone" (*lapis philosophorum*) that in the best-known stories was a substance that could convert base metals into silver and gold. But as with almost everything else in the hermetic tradition, the story is much more complicated than that. Like the Grail legend, with the Philosopher's Stone it can be difficult to separate the literal from the allegorical. Thus, the stone was also believed to be the "elixir of life," capable of staving off mortality for centuries. It was also the symbol of enlightenment.

The search for the Philosopher's Stone was part of a larger quest called the "Great Work" (which, by the way, also included the Grail). In the words of the nineteenth-century French occultist Eliphas Lévi:

> *The Great Work is, before all things, the creation of man by himself, that is to say, the full and entire conquest of his faculties and his future; it is especially the perfect emancipation of his will.*[2]

Though its practitioners might not agree, beyond the obvious attraction of gaining divine knowledge and its attendant power, part of the appeal of the Great Work was the sheer impossibly of achieving its goals. The quest itself had its own cultural power—it could even get you burned at the stake . . . and a statue raised to your memory.

The irony of this five-thousand-year quest was that, even as each generation of occultists spent their lives in fruitless search, the path to this infinite knowledge was being forged, inch by inch, by the least likely (and least mystical) of explorers: scribes and printers, tinkerers and engineers. The difference between these two groups of questers could not have been starker, and that difference was no more apparent than in 1969.

To have believed the media coverage that year, it was a turning point in human history. At Woodstock, the counterculture had its coming-out party, celebrating the new power of youth and (it was said) ushering in a new age of love and enlightenment. And as *Apollo 11* landed on the moon, the world rejoiced at man's first giant leap into space.

Yet, with the hindsight of decades, as the baby boomer generation grew old and NASA, having stopped visiting other worlds in 1972, eventually abandoned even the Space Shuttle program, it was apparent that this new age was over almost as soon as it had begun.

Meanwhile, men with crew cuts instead of shoulder-length hair, white shirts and skinny ties instead of tie-dye, and lab coats instead of spacesuits were buried in laboratories and offices creating a real point of inflection in the story of the human race. We can see now that it was the invention of the microprocessor and the creation of the Internet that made 1969 a true year of miracles.

After millennia of continuous improvement and innovation in the gathering, preservation, organization, and presentation of memory, these two breakthroughs and all of the many inventions of the digital age that supported them—magnetic memory, computers, networks, displays, and so on—had created a wholly new and unexpected kind of Philosopher's Stone, a vast global Aleph of memory.

In the late 1990s, as the implications of the World Wide Web became more clear, a prophesy in the form of a thought problem briefly circulated in Silicon Valley. It asked:

> *What if you had a small box—an Answer Box—that contained all of the world's knowledge and memories? No matter what question you asked it, it would not only provide the answer but present it in any way you wanted it—audio, video, tactile—directly into your brain. What would you ask it?*

Behind that question was the implication that now that this Answer Box, the dream of mankind—almost since mankind *could* dream—was seemingly within our reach, had we prepared ourselves for it? And if not, could there be a greater tragedy than to have all the answers waiting for us . . . and not be able to formulate the right questions?

Had humanity at last built a machine that was beyond our capacity to use it?

And would the easy availability of knowledge and memory cheapen its perceived value?

Those were not comfortable questions to ask, nor easy ones to answer, and the Answer Box paradox disappeared as quickly as it appeared. But the problem still remained, and thanks to Moore's Law, it grew closer by the year. And the search for the answers to those questions has only been postponed. Once we take them up again, they will inevitably lead us back to where we began: the human brain and its capabilities.

After the ancient art of memory enjoyed its revival during the Renaissance—and was generally considered a failure—the study of human memory faded in importance to the occasional and anecdotal (the savant, the rare individual with a photographic memory, amnesia cases) as progress in artificial memory proceeded at breakneck speed. After all, why spend years perfecting memorization techniques when books were becoming cheap enough to fill a middle-class home library—and one could access thousands of volumes in the growing number of free, public libraries?

Not surprisingly, the memorization of key texts, which had been a centerpiece of education not only in the ancient world but in the Middle Ages and Renaissance, slowly faded from the curriculum. By the Enlightenment, rote memorization not only seemed a sign of intellectual rigidity but also a waste of time that could be better spent reading more books. Our grandparents, parents, and even many of us in our youth, in what we now often think of as repressive classroom environments, were required to know pieces of the text of a few historic documents (for example, the preamble to the U.S. Constitution), speeches (the Gettysburg Address), songs, and poems ("Paul Revere's Ride," "The Charge of the Light Brigade"). Today, in most schools in the developed world, even that little bit of memory work is gone, leaving only the memorization of a few mathematical and scientific equations and perhaps the lines to a part in a school play—and even that is seen as onerous.

The typical modern school test is often taken either with open notes and textbook or with a calculator. And why not? Memory is now free, ubiquitous, and almost infinite; what matters now is not one's ownership of knowledge but one's skill at accessing it and analyzing it. The last great argument for memorization—that is, what would you do if you found yourself in a situation without a calculator or the right manual?—became almost meaningless in a world when both were now readily available online anywhere on the planet, from the Serengeti to Antarctica.

What value remained in one's private memory no longer came from what might be called "common" knowledge; it was usually far more accurate to search the Web than one's own memories for information about episodes from old television shows, the lyrics of hit songs, and the precise chronology of past events than it was to consult one's own incomplete and biased personal memories. Indeed, it often seemed that the only "brain memories" that still really mattered were those that were intensely personal. By the twenty-first century, as the Web, security cameras, and social networking sites increasingly made the even the most intimately personal into a shared public experience, it began to seem that the only important personal, biological memories that had value were those so quotidian, small, and inconsequential that the rest of the world simply wouldn't be interested. Francis Bacon was still right: Knowledge—memory—was still power, but it wasn't our knowledge or our memory. In the world of microprocessors and servers, social networks and the World Wide Web, power was now *access* to the most valuable caches of memory.

GRAY MATTERS

Ironically, even as the value of the individual human brain diminished, the understanding of the power and complexity of that brain was increasingly understood thanks to the rise of experimental science. Physicians and scientists throughout the nineteenth century, working mostly with stroke victims, the mentally ill, and brain-damaged war veterans had slowly begun to piece together a model of the human brain and a map of its various functions. Then, as now, these researchers were most haunted by amnesiacs—otherwise normal people who had (temporarily or permanently) lost all of their accumulated memories and found themselves in the living hell of being without a past and without an identity.

In the last decades of that century, the Austrian neurologist Sigmund Freud, working from an idea first proposed by the German philosopher Theodor Lipps, began to study the functional operation of the human brain through a process of deep conversation and dream analysis—psychotherapy—with his psychologically troubled patients. What Freud discovered, and what made him one of the most influential scientific forces in the coming century, was that the brain, whatever its underlying physiological structure, in action was an incredibly complex organism that operated at least as much below the surface of consciousness as above. And it was this "subconscious," often containing memories so embarrassing or traumatic that the brain had repressed them into this hidden location, that continued to secretly work its damage on a person's behavior.

Carl Jung, Freud's erstwhile colleague, looked at this same unconscious and believed he saw hidden memories—"archetypes"—that seemed to be common to all mankind past and present, and with each of us from birth. Jung suggested that this "collective unconscious" might represent a very primitive sort of universal mind that might have superhuman powers—a notion largely dismissed as yet the latest eruption of hermeticism. That is, until the rise of the Internet.

In the first half of the twentieth century, even as the general public was assimilating Freud's and Jung's theories, brain research moved from the therapist's couch into the laboratory. There, scientists such as the Russian Ivan Pavlov in the 1920s and the American B. F. Skinner in the 1930s studied how the mind learns behavior by repeatedly accessing memories hidden in the unconscious.

By the middle of the twentieth century, thanks to a whole spectrum of new medical analytic tools made possible by the digital revolution, scientists were increasingly able not only to probe the structure of the brain through targeted X-ray and magnetic resonance imaging but, through the tracking of electrical stimulation, to actually see the brain in action. The result, beginning in the 1960s and continuing to this day, is an increasingly sophisticated and nuanced model of a brain that is anything but simple and monolithic. Here, in summary, is what we now know:

The average human brain weighs about 1.5 kilograms (3 pounds) and has a volume of about 1,200 cc. Brain size is related to body size, so male brains are typically about 100 cc. larger than female brains; and while in extreme cases brain size can be indicative of severe retardation, in normal brains there is little correlation between size and intelligence.

Structurally, the human brain contains just over 200 billion nerve cells. Half of these are *glial* cells, which provide support for an equal number of *neurons*, the latter doing the work of thinking. In most of the brain, these glial cells are teamed one to one with neurons, acting as everything from insulators to transmission managers; in the upper brain, the "gray matter" of the cerebrum, that ratio is one to two. The cerebrum also contains 10 billion high-performance *pyramidal* neurons.

Unlike computers, where the transistors in chip memory and the magnetic bit locations in disk memory are arranged basically in a linear manner, animal brain neurons have connectors (*ganglions*) arrayed like the roots and branches of a tree that connect with the similar arrays of numerous other nearby ganglions—connections that are strengthened with use. This *multiplexing* enables the average human brain to exhibit as many as 1,000 trillion—fifty quadrillion—connections. You'll notice this means that just fifty thousand people have as many brain connections as there are total bytes on the global Internet.

The brain itself consists of several large regions. The main mass of the brain consists of two mirror-image hemispheres, themselves consisting of the "white" (or light-gray) matter of the basic mammal brain; the *cerebrum*, resting atop the *brain stem* which connects to the spinal cord; and in the back at the bottom is the *cerebellum,* whose furrowed surface resembles twisted rope. The cerebrum manages the basic mental operations of the brain; the cerebellum, the direct descendant of the brain of older animal phyla, manages the body's motor functions; and the brain stem carries messages to and from the body's muscles, organs, and glands to and from the brain.

The cerebrum itself is covered by a comparatively thin, but heavily convoluted (to increase surface area), *cerebral cortex*. Roughly speaking, the more intelligent the animal, the more convoluted its cortex, with man having the most "wrinkled" brain of all. The cerebral cortex, as noted earlier, is surprisingly large when unwrinkled and laid out flat—more than 2.5 square feet. And it directs the higher thinking found mostly in primates. In human beings that includes speech, language, logical thinking, vision, fine motor skills, metaphor, analogy, and so on.

To simplify matters, the cerebral cortex is usually divided into four general regions—"lobes"—on each hemisphere and named for the skull bones that encase them: *frontal* (ambition, reward, attention, planning, and short-term memory tasks); *parietal* in the top back (ties together sensory information relating to spatial sense and movement); *occipital* in the far back (vision); and *temporal* on the lower sides (hearing and speech).[3]

THE GEOGRAPHY OF MEMORY

In light of the narrative of this book, the obvious question to ask at this point is: Where does memory fit in all of this?

The answer, researchers have found, is that it fits almost everywhere. Memories appear to be stored throughout the brain in a manner, and according to rules, that have yet to be fully explained. Moreover, as anyone who has ever tried to dial a telephone number after just hearing it, or crammed for an exam, or suddenly remembered some trivial detail out of a far-distant past, human memory is not a monolithic process.

In fact, neurologists have identified three primary memory activities and three primary memory types. The first three are implicit to the nature of memory itself, and thus can be found in both human and artificial memory: *encoding,* the capture and preparation of information for preservation; *storage,* the recording and archiving of that information; and *retrieval*, the locating and removal of that information from storage.

But the architecture and form of the organic human brain is very different from artificial computer memory. Though there is a superficial similarity between cache, ROM, and RAM and what scientists call the brain's *sensory, short-term*, and *long-term* memory, they have radically different purposes and causes. In the computer, cache memory is essentially a waiting room for processing, ROM is the home of operating tools that are protected from modification, and RAM is a vast warehouse of undifferentiated memory denoted only by address.

By comparison, the brain's sensory memory—the ability to capture and hold on to an enormous amount of information taken in by the senses in what has been determined to be less than a half-second—appears to be a genetic response to the complicated natural world. That is, to "see" more than you actually see in case it is hiding prey . . . or a threat. Tests have found that human beings can capture up to twelve items at a glance . . . but forget most of them in less than a second. Importantly, it seems that it is impossible to improve the direction of one's sensory memory with practice.

Short-term memory, as already noted, is typically stored in the frontal lobe. It has its own limitations—as anyone knows who has tried to hang on to a name or address from the time you hear it until you try to write it down even a moment later—especially when there is even the slightest interruption.

In 1956, George Miller, a cognitive scientist working at Bell Labs at the same time as William Shockley, published one of the most cited papers in the history of psychology. Entitled "The Magical Number Seven, Plus or Minus Two," it made the case, based on studies with test subjects asked to remember lists of words, numbers, letters, and images, that the human brain was able to briefly—meaning up to a minute without rehearsing—remember about seven items on a list, plus or minus two items. Later research has put that number closer to the lower end of that range.[4]

There are some tricks to increasing both the size of short-term memory and the duration of its storage. The first, as noted earlier in this book, is "chunking," which takes advantage of the brain's ability to treat small clusters of information (usually no more than three items) as a single chunk of memory, which is why humans can often remember a phone number better by breaking it up (in the United States) into the area code and local prefix—each three numbers—and then the final four digits into two-number pairs.

As for duration, the solution, as every student knows, is repetition. Short-term memory appears to be a largely chemical process that fades quickly. Thus, if that memory can be quickly pumped up again to a full charge before it disappears—and this process is repeated continuously—information can be retained in short-term memory for an extended period. Better yet, the constant reinforcement of short-term memory seems to be the brain's primary criteria for transferring that information into long-term memory.

As for long-term memory, it is a whole different creature indeed. What makes it astonishing is that, at least by a human scale, it seems both infinite and immortal. For example, there seems to be almost no limit to the number of memories that the human brain can hold—remember that massive number of connections. It is possible that *every* memory you ever experienced that made its way into your long-term memory is still buried somewhere in your head, and it is just the insufficiently powerful catalog and search tools in your brain that keeps you from finding them. We've all had the experience of concentrating on remembering something, then giving up . . . only to have the answer pop into our minds hours, even days later, suggesting that the search took longer than we expected. By the same token, all of us have had the experience of thinking about something . . . only to have some completely different long-forgotten experience or memory pop into our minds, suggesting that it was accidentally captured along with an adjacent, targeted memory.

By the same token, once an item is stored in long-term memory, it seems to last forever unless it is in some way destroyed by injury, disease, or death. A memory from the crib, if strong enough to persist, can be remembered a century later by an aged centenarian as vividly as the day it was forged. Were we to suddenly live five hundred years, there is no reason that same memory from a half-millennium before wouldn't still be fresh and bright.

OTHER MINDS, OTHER MEMORIES

It's an extraordinary organ, the human brain and its memory. It doesn't seem so absurd now that when the ancients attempted to take memory to a higher level, they chose to pursue that goal internally and organically rather than externally and artificially. That they failed doesn't diminish their attempt; rather, just imagine how the different course of human history would have been had they succeeded.

But they did fail. And for thousands of years, we have pursued a different path; one that is outside of our skulls and that, for all of its power, must forever find a way back inside, with all of the associated compromises of access and translation.

Now, after all of the intervening centuries, the two paths seem again to be converging.

In recent years, machines, especially those that have jumped aboard the rocket of Moore's Law, are achieving a level of raw intelligence that approaches—and in some cases even exceeds—that of the human brain. At the same time, this artificial intelligence is spreading far from its traditional home in computational devices and test-and-measurement instruments to every corner of daily life. And that includes sensors, pattern-recognition devices, vision systems, nanomachines, and hundreds of other technologies that lend themselves to supporting—as the human body does with the human brain—the interconnection of computer intelligence and the natural world. Most of these peripheral devices exhibit performance well beyond that found in even the most proficient human beings.

There is another factor as well: Compared to the human brain, these digital devices are also breathtakingly fast. The basic clock of the animal world is the heartbeat, and it is a rule of thumb that most living things have within them about 10^9 (1 billion) heartbeats. Thus, animals with rapid heartbeats (insects) have short lives; those with comparatively slow heart rates (primates, tortoises, parrots) have long ones. By comparison, as this is being written, modern state-of-the-art microprocessors have clock speeds approaching 5 gigahertz—or 5 billion cycles per second. In other words, these chips—and the devices they run—experience the equivalent of several human "lives" every second.

Finally, to this mix add the Internet, by many orders of magnitude the largest repository of memory ever created—and optimized for navigation by computer intelligence. Indeed, unlike a library, the World Wide Web is a place that can only be entered accompanied by a computer or other digital device.

Once again, memory is power. And the history of humanity can been seen as the long, long story of the increasing distribution of the ownership of memory—and thus liberty—from the few to the many, from shamans and kings to everyone, including the most wretched of mankind. Memory liberates, so are the next subjects of that liberation our machines? After all, they now control most of the world's memories.

None of these questions have been lost on humanity. On the contrary, in the two centuries since Mary Shelley's *Frankenstein*, and especially since the rise of science fiction in the twentieth century, we have been increasingly obsessed with the idea of intelligent machines—at best as our loyal compatriots, at worst as our evil overlords. And if we ponder that story about the Answer Box long enough, it becomes apparent that the real problem may not be what question we would ask it, but whether it needs us to do the asking at all.

This is not to suggest that a world of humans and conscious, independent machines casually interacting on an everyday basis—or worse, a dystopian world in which humans are enslaved by far superior silicon-based life forms—is anywhere in our near (or distant) future. However, the increasing convergence of the two forms of memory, natural and artificial, suggests that some kind of reckoning lies just a generation or two ahead.

MASTER, PARTNER, SELF

What will this reckoning look like? There are three likely scenarios: living machines, assisting machines, and human machines.

Living Machines

As we've seen, human beings have been trying to make their creations look and act like living things ever since the ancient Greeks and the Chinese. But it was de Vaucanson and the automaticists who first created mechanical devices that could effectively mimic a wide range of animal and human behaviors. But imitation isn't actuality, and no matter how stunningly real the Flute Player or the Digesting Duck might seem in a controlled setting, its repertoire was small; and no matter how great the skill of the builder, none of the creations ever exhibited any of the traits we think of as being alive, from reproduction to adaptability to their surroundings to self-maintenance. They did the same tricks over and over until they broke, all the while looking out at the world with dead eyes.

Though they failed at their immediate purpose, the immense influence of these automatons has been almost immeasurable. Not only, as noted, were their toothed-wheel controllers an important marker on the path to modern computing, but their actual gear, axle, and pulley mechanisms were crucial to the development of twentieth-century cybernetics. When, in the 1920s, playwrights like Karel Čapek (with his humanlike androids in the play *R.U.R* that coined the word "robot"), filmmakers like Fritz Lang (with *Metropolis* and its beautiful *Maschinenmensch*), and inventors like biologist Makoto Nishimura (with his robotic bust of *Gakutensoku* that could laugh and cry and turn its head) began to create the image of the modern robot, it was that of an automaton with a tabulating machine for a brain.

The assumption was, in a kind of presaging of Moore's Law, that it was only a matter of time until the mechanical systems governing the motion of robots were sufficiently precise and reliable that they would be, to the naked eye, all but indistinguishable from living organisms. By the same token, as the tabulating machines became computers, and the computers developed the complexity of organic brains, it was also believed that robots would also begin to "think" like living things . . . and eventually "wake up" to a kind of servile consciousness.

In the end, we got most of the first and not much of the second. Today's robots, especially those created in university laboratories, do a very good job of re-creating bipedal motion, or carefully picking up objects, identifying unique patterns, recognizing spoken words (especially from a single speaker), and constructing verbal sentences in an intelligible voice. But as impressive as these constructions may be, they are still disconcertingly far from truly autonomous creatures. And worse, the closer they come to achieving their goals, the deeper they seem to sink into what has been called the "uncanny valley," in which the more lifelike an artificial form becomes the *less* alive it seems. Thus, Mickey Mouse still seems more real to us than the soulless creatures created by the latest computer graphics programs to look almost identical to real human beings on-screen.

This quest to re-create life in an artificial form has taken a backseat to the real business of modern robotics: the construction of mechanical slaves to take on tasks that are too dangerous or repetitive to still be done by human beings—wrapping wiring harnesses, welding truck quarter panels, picking up newly cut integrated circuit chips and welding interconnects, grabbing items from sea beds, and, increasingly, performing tasks in surgery and dentistry. These robots are, for the most part, fractional entities—giant arms, precise fingers, motorized tracks following buried wires. They are also single-minded in their purpose; most are programmed via a local-area network and have little or no contact with the Internet. If we wait for these machines to "wake up," we may wait forever.

But what of the big multiprocessing supercomputers? They certainly match or surpass the human brain in many areas of performance, including processing speed. Will they begin to think autonomously sometime soon, establish their own identity, and achieve some kind of will and consciousness? Predictions of big computers thinking on their own—perhaps even exerting control over mere mortals—are as old as mainframes themselves, and seem to gain new speculative life with every new generation of "Big Iron."

And yet, other than a few anecdotes—the best known being the famous 1996 and 1997 matches between world chess champion Garry Kasparov and the IBM Deep Blue supercomputer, after which Kasparov said he sensed a mind at work in his opponent—there is no indication that a computer has ever, even for a second, accomplished "thought" as we conceive of it in living things, much less achieved a consciousness of its own existence.

That could change someday—perhaps sooner than we think. The Blue Brain Project, begun in 2005 at the Swiss École Polytechnique, is using an IBM supercomputer to replicate the actual mammalian brain, right down to the structure of its neurons. Speaking just yards from the Bodleian Library at Oxford, Blue Brain director Henry Markram announced, "It is not impossible to build a human brain, and we can do it in ten years"[5] To the BBC he added, "If we build it correctly it should speak and have an intelligence and behave very much as a human does."[6]

Time will tell. And what of the Internet itself? With its wireless and dial-up links mixed in with its ultrabroadband trunk lines, it is slower than the human brain but a thousand times more powerful, and it features much of the same multiplexing that is found in animal neurons. Does the

Internet think? And if so, and it becomes, as H. G. Wells predicted in 1938, "a world brain," will it have at its command all of human memory and knowledge? Will we really, as Wells claimed, embrace it because "we do not want dictators, we don't want oligarchic parties or class rule, we want a widespread world intelligence conscious of itself"?[7]

Perhaps—and perhaps not. But it is hard not to dispute the prescience of the rest of Wells's prediction:

> *The whole human memory can be, and probably in a short time will be, made accessible to every individual. . . . This new all-human cerebrum need not be concentrated in any one single place. It need not be vulnerable as a human head or a human heart is vulnerable. It can be reproduced exactly and fully, in Peru, China, Iceland, Central Africa, or wherever else seems to afford an insurance against danger and interruption. It can have at once, the concentration of a craniate animal and the diffused vitality of an amoeba.*[8]

In 1997, George Dyson, son of the noted physicist Freeman Dyson, published *Darwin Among the Machines.* In it he looked positively upon the idea of sharing the world with intelligent machines and warmly anticipated what he thought to be their impending arrival. He approvingly quotes the essayist Garet Garrett, who wrote in 1926:

> *Man's further task is Jovian. That is to learn how best to live with these powerful creatures of his mind, how to give their fecundity a law and their functions a rhythm, how not to employ them in error against himself.*[9]

Dyson then asked: "Is the diffusion of intelligence among machines any more or less frightening? Would we rather share our world with mindless or minded machines?"[10] For George Dyson, the answer was clear: Artificial intelligence, even consciousness, was inevitable, and by the right of successful evolution (even if it was by man himself), machines had to be allowed to fulfill their own destiny. As Dyson put it, "We are brothers and sisters with our machines . . . in the game of life and evolution there are three players at the table: human beings, nature and machines. I am firmly on the side of nature. But nature, I suspect, is on the side of the machines."[11]

But if the Internet was ever going to awaken, like a great digital Leviathan (to use one of Dyson's favorite analogies), and embrace that destiny, it probably should have begun stirring by now. And yet . . . nothing.

Or more accurately, nothing *yet*.

INSIDE JOB

Assisting Machines

In 2011, teacher Michael Chorost published *World Wide Mind: The Coming Integration of Humanity, Machines, and the Internet*. Its subject was in its title, but the implicit message of the book was: Why wait for our machines to awaken and come to us? Instead, let's meet them halfway. . . .

> *What if we built an electronic corpus collosum [the linkage between the two hemispheres of the brain] to bind us together? What if we eliminated the interface problem—the slow keyboards, the sore fingers, the tiny screens, the clumsiness of point-and-click—by directly linking the Internet to the human brain? It would become seamlessly part of us, as natural and simple to use as our own hands.*[12]

In particular, what Chorost proposed was that, using a complex technique combining viruses to alter the DNA of brain neurons, optogenetics (using light to control cell functions), and the implantation of nanowiring, it should be possible to install a wireless modem directly into the human brain. The process would be difficult and time consuming, Chorost admitted, and the time needed to actually control this new part of the brain might run into months . . . but in the end, the owner of this modified brain would be able to communicate with other human beings with similar brains in a manner not unlike telepathy, or at least like people instant messaging each other with their cell phones.

But Chorost went one step further. He argued that if all human beings had their brains modified in this way, they would be able to link together in a vast mental network—a "hive mind" is the term applied to social insects—that would be greater than the sum of its parts, and would result in deeper human relationships, larger and more successful group endeavors, and greater mutual understanding.

Chorost came with unique credentials. Born nearly deaf from a case of rubella, he continued to lose the rest of his hearing into adulthood. Finally, in 2001, unable to get by with just hearing aids, he underwent the still-experimental surgery of a cochlear implant—a device that combines a microphone, a digital speech processor, and transmitter—to capture and filter sound and then transfer it directly from the inner ear to the auditory nerve to the brain.

From that life-transforming experience, Chorost wrote *Rebuilt: How Becoming Part Computer Made Me More Human* in 2005. His yearlong experience in learning how to hear again, and the transformative effect this restored sense had on his life, led Chorost to investigate what it would take to insert even more powerful technology into the human brain.

To date, about a quarter-million people have received cochlear implants. Thousands more have received deep-brain implants for brain and vagus nerve stimulation to help them fend off the effects of Parkinson's disease and depression. Others have received brain "pacemakers" to manage epilepsy. Others are presenting themselves as test patients for miniature camera brain-implant systems to restore sight.

These brain implants seem to point the way toward even more transformative uses—not just Chorost's dream of a World Wide Mind, but something more personal and individual: the ability to add new memories, knowledge, skills, and talents directly into the human brain from artificial sources.

Brain implants have been around for a surprisingly long time. As early as 1870, German researchers Eduard Hitzig and Gustav Fritsch had implanted an electrode into parts of a dog's brain and stimulated it to repeat certain movements—a technique that was reproduced in the human brain. In fact, what we now know as the map of the human brain was largely discovered through the use of these implants.

By the mid-twentieth century this implantation mapping technique had become very sophisticated and capable of identifying and diagnosing certain forms of mental illness. But the discipline really took off with the arrival of computers, magnetic-resonance imaging, and three-dimensional imaging; now brain function could not only be statically mapped but also studied in real time as different regions lit up with electrons in use. Indeed, it was even eventually possible to make fairly accurate guesses what patients were thinking about given the unique pattern of their firing brain neurons.

In one of the most remarkable studies, undertaken in 1999 by a team at the University of California–Berkeley, researchers implanted 177 electrodes into the thalamus region of a cat's brain (the part of the brain that translates sensory data into brain signals)—in particular the part connected to the optic nerve. They then tracked the firing neurons in the cat's thalamus and ran the results through a computer using a process called linear decoding. They were astonished by what they saw. It was the world as seen through a cat's eyes, including their own faces.[13]

The next decade saw significant—and often controversial—progress in this emerging field of *thought identification*. Using a new brain-scanning technology called functional magnetic resonance imaging (fMRI) to track the changes in blood flow resulting from neural activity, researchers were able to do what was heretofore considered magic: to predict human action, such as pressing a lever, *before* a subject knew he or she would do it—that is, reading the unconscious brain making a decision (and the decision it would make) before that choice ever reached the conscious mind.

This predicting of human intention was controversial enough, with the accompanying major ethical concerns regarding privacy and philosophical implications about free will. But the researchers had just begun. The discovery that all human brains respond to the same images in the same way meant that numerous images could be shown to subjects, the fMRI patterns tracked and cataloged in computers, and a vast encyclopedia of brain images created and then accessed in real time. And that, in turn, made it possible to "read" a patient's thoughts and memories in real time.

By 2007, Barbara Sahakian, a professor of neuropsychology at Cambridge University, was able to say with (chilling) confidence: "A lot of neuroscientists in the field are very cautious and say we can't talk about reading individuals' minds, and right now that is very true, but we're moving ahead so rapidly, it's not going to be that long before we will be able to tell whether someone's making up a story, or whether someone intended to do a crime with a certain degree of certainty."[14]

So that's the *reading* of thoughts and memories in the brain. What about the *writing* of experiences directly into the brain?

The idea of manipulating memory by putting thoughts (usually false) into the brains of others is at least as old as Descartes. A corollary to his process of stripping away all indisputable knowledge from his brain to reach the only surviving truth—that of his own thinking as proof of his existence (*cogito ergo sum*)—was the possibility that all of his other memories and observations might not just be untrue, but *intentionally* false. Descartes imagined it as his brain in a black box with an evil demon controlling everything going into and out of that box. That was 1638, but it remains a notion as current today as the *Matrix* movie trilogy.

"Brainwashing," the psychological technique of inserting false memories into others, burst into the public eye in the early 1950s during the Korean War, when North Korean interrogators were accused of using the technique to distort the psyches of captured U.S. soldiers. It was a process made vivid by the movie *The Manchurian Candidate*. It surfaced again during the sex-abuse hysteria of the 1980s, when a number of child day-care-center operators (notably the Amirault family in Malden, Massachusetts) were accused—through the "recovered memories" of young children—of administering bizarre sexual rites on the children in their care. Ultimately, the charges against these operators were dismissed when it was determined that prosecutors had systematically convinced those children to believe, without evidence, the truth of these claims.

But the idea of implanting empowering, rather than destructive, memories into the human brain first really captured the public's imagination with the publication of the first "cyberpunk" science-fiction novel, *Neuromancer*, by William Gibson, in 1984. The world of *Neuromancer* is filled with characters, mostly mercenaries, who regularly enhance their performance—memory, strength, sight, skill sets—by "jacking in" brain implants to download this knowledge from the universal "matrix." Gibson's literary skill, combined with his extraordinary ability to extrapolate current technology into the future (via a Moore's Law–like view of modern life), almost instantly made the notion of brain implants not only possible but something to be anticipated in the near future.

Unfortunately (or perhaps fortunately), it hasn't quite turned out that way. In theory, placing digital technology into the brain should be easy. In reality, it has proven to be very difficult. Note how complex Michael Chorost's model of the brain modem was to execute, requiring genetically engineered viruses (the alternative being dangerous open-skull brain surgery and the precise locating of microscopic wires and other devices). Chorost's cochlear implant, like other sensory restoration techniques, is comparatively easy because it mostly sits on the *outside* of the brain, or even the skull, not inside the brain itself.

That's why reading the brain is so much easier than writing on it. You can slice the brain with fMRI and watch it work, but you can't do that continuously day after day without risking injury. By the same token, you can wrap the skull with a cap embedded with scores of electrodes—and leave it on permanently—but now you can only read the operations of the brain within, and not very precisely.

In other words, even reading the brain is tough, and it requires serious compromises between precision and permanence. Writing on the brain is far tougher. Experimental brain implants have proven to be very precise and powerful about turning on and off different actions and memories. But to electronically implant a *new* memory is not yet possible, and is many magnitudes greater in difficulty—after all, the brain scatters memories all over the place. And even then, there are serious questions about how long the apparatus will work. The brain is a living organism and shares the body's immune system, and past experience with electrodes has found they begin to fail after a few weeks as the brain begins to surround them with scar tissue. Even if we could selectively turn off the immune system in the brain to keep these implants functional, we'd then be opening the door to infection and possibly cancer.

So, as thrilling as the notion is of having a small slot in the back of one's head where you can swap in and out a knowledge of French, automobile repair, or American history as easily as you can a memory card into a digital camera or a thumb drive into a laptop computer, that reality is a long ways off. And even if there were to be a breakthrough in machine-brain interfacing, it is not self-evident just how all of that new memory would be used. Would it all be poured into the brain at once, or would the brain regularly talk to the memory card? And just what kind of translating of this data and training of the brain would be needed to make it work?

Based on his own experience of needing months to make full use of his cochlear implant—and he wasn't fully deaf from birth but had experience actually hearing—Michael Chorost suggests that learning to use his brain modem might take a year or more. That's a lot of pain and commitment for a result that might soon fade away. The obvious solution, assuming all of the other obstacles are overcome, would be to embed the "memory plug" into a newborn—or, better yet, a fetus. But in a world where removing a baby boy's foreskin is becoming criminalized, what is the likelihood of legally putting a plug into a newborn's brain?

However, as unpleasant and painful as brain implants may sound, we can still assume that if the technology even approaches practical implementation there will be a small army of volunteers willing to take on the misery and risk of being a pioneer of the brain memory implant. Since the turn of the millennium, as ever greater numbers of people have received not just brain-oriented devices such as cochlear implants but also artificial limbs, bones, tools for locomotion, and so on, a cultural movement—*transhumanism*—has emerged that is dedicated not just to the advancement of human-machine technology, but its celebration. Transhumanists see the use of machines and computers not as a last-ditch effort to restore a failed biological system but rather as a means to enhance human existence. In their vision, at George Dyson's table there will one day be only two players present: nature and a hybrid of man and machine . . . and Mother Nature will learn to love her flesh-and-metal child.

A SINGULAR TURN OF EVENTS

Human Machines

On the farthest shores of this new world of artificial/natural memory lies the strangest vision of all, one to which many of the transhumanists aspire . . . and one that some of the world's most brilliant computer scientists believe is both inevitable and immanent.

It is called the *Singularity*.

The term "Singularity" has a lot of different definitions in mathematics, cosmology, and quantum physics, but all share a common attribute of characterizing a moment or location or event where everything undergoes a massive change so complete that comparing before and after is almost a meaningless exercise. The same is true for what has been predicted to be a Singularity in technology: It will transform the meaning of what it is to be human or a machine, of natural and artificial memory, life and death, and ignorance and knowledge so completely that, from this side of that event it is literally impossible to predict what will take place on the other side.

Compared to the other two scenarios, the Singularity is a relatively new concept. Presaged in the 1960s by any number of movies, TV episodes, and science-fiction stories predicting a future in which computers and robots suddenly break through an invisible barrier and become self-aware, and self-improving at a blinding speed, the idea of the Singularity was first proposed by the British statistician (and one of Alan Turing's old compatriots) Irving Good in 1965:

> *Let an ultraintelligent machine be defined as a machine that can far surpass all the intellectual activities of any man however clever. Since the design of machines is one of these intellectual activities, an ultraintelligent machine could design even better machines; there would then unquestionably be an "intelligence explosion," and the intelligence of man would be left far behind. Thus the first ultraintelligent machine is the last invention that man need ever make.*[15]

The Singularity itself received its first formal definition in a 1995 essay entitled "The Coming Technological Singularity: How to Survive in the Post-Human Era" by mathematics professor and science-fiction writer Vernor Vinge.[16]

The title says it all. For Vinge, the Singularity would be an explosion of artificial intelligence that would ultimately result in a "superintelligence" whose transformation of the future would be as complete and inexplicable as the event horizon on the edge of a cosmic black hole. Thus, the first of these superintelligent machines will be mankind's last and greatest invention. After that, humanity will be largely superfluous.

Vinge's most famous quote about this Singularity is: "Within thirty years, we will have the technological means to create superhuman intelligence. Shortly after, the human era will be ended."[17]

In Vinge's scenario, at the Singularity one or more of our computers (or, pace Dyson, the Internet itself) becomes so intelligent and competent that it starts re-creating and upgrading itself or begins building other computers even more capable than itself . . . and in no time, mankind is left in the dust. Or, in the most dystopian case, we become slaves to our new digital masters. In fact, when you begin with Vinge's Singularity, things can get really ugly really fast. For example, consider this nasty little example of the Law of Unintended Consequences at the Singularity, courtesy of Nick Bostrom, philosopher and director of the Future of Humanity Institute, which just happens to be across the street from the Bodleian Library:

> *When we create the first superintelligent entity, we might make a mistake and give it goals that lead it to annihilate humankind, assuming its enormous intellectual advantage gives it the power to do so. For example, we could mistakenly elevate a subgoal to the status of a supergoal. We tell it to solve a mathematical problem, and it complies by turning all the matter in the solar system into a giant calculating device, in the process killing the person who asked the question.*[18]

Hold on, says Ray Kurzweil, the renowned inventor and the figure currently most associated with the Singularity. Why does mankind need to be left behind on this side of the Singularity? Why can't we go forward as part of our machines?

Kurzweil, who was raised in Queens, New York, first learned about computers from his uncle, who was an engineer at Bell Labs. In 1963, at just fifteen, Kurzweil wrote his first computer program and in short order was winning national and international science fairs with his inventions. Over the next thirty years, Kurzweil made his reputation with one invention after another, developing tools that enabled the blind to read, synthesizers that finally reproduced the sound of traditional instruments, speech-recognition computers, and virtual-reality training tools for medical professionals.

Then, beginning in 1990, Kurzweil embarked on a series of three books of predictions about the technological future. These books, and the theories they present, have dominated Kurzweil's career ever since—and given him a reputation as one of the world's leading futurists. At the heart of all of these works is Kurzweil's belief in Moore's Law as not only the defining force of our time but the most powerful tool available to predict the new world—in other words, the Singularity—that Kurzweil believes (and has convinced millions of others to believe) is likely to arrive within most of our lifetimes.

The titles of the three books—*The Age of Intelligent Machines* (1990), *The Age of Spiritual Machines* (1999), and *The Singularity Is Near* (2005)—show both the development, and the increasing optimism, of Kurzweil's thinking. *Intelligent Machines* was essentially an extrapolation using Moore's Law from the existing pre-Internet technology of the late 1980s into the decade or two ahead. Though some of his later claims about his predictions are a bit far-fetched (such as having predicted in 1986, when he began the book, the impending fall of the Soviet Union) he proved as prescient as anyone in anticipating the explosion not just of the Web, but also of wireless telecommunications. He also predicted the future defeat of a human chess champion by a computer program (which took place in 1997 with Kasparov's defeat by Deep Blue).

With *Spiritual Machines*, Kurzweil grows much more ambitious in his predictions and his time horizons. The book is largely cast in the form of a conversation between Kurzweil and "Molly," a fictional foil. The book begins in 1999 with Molly as an average young woman, largely uninformed about the information revolution and a little flirty. The book ends four hundred pages and a century later, with Molly now evolved into a conscious, noncorporeal self embedded within a powerful computer. She is now brilliant, curious, and ready to experience the entire world.

PLACING BETS

Part of the fun (and the courage) of *Spiritual Machines* comes at the end, where Kurzweil makes predictions, by the decade, for the century ahead. With the first set, for 2009, he offered 108 different predictions—the rise of portable computing; the decline of disk drives; movies, books and music increasingly being delivered digitally; tele-medicine; and so on. Time has proven him to be impressively accurate, though more with his technological than cultural and economic predictions.

Looking further into the future, Kurzweil's predictions for the end of the twenty-first century are pretty radical: thousand-dollar computers as powerful as the human brain by 2019; food commonly assembled by nanomachines by 2049. By 2099, Kurzweil sees a world in which artificial intelligence is not only superior to human intelligence but one in which AI dominates the landscape, with humans embedding implants in their brain just for the chance to be part of the AI world . . . and the remaining "traditional" humans protected by machines like exotic wildlife. Meanwhile, those humans who have turned themselves into artificial life forms regularly make backup copies of themselves to obtain a kind of immortality.

Kurzweil finishes his predictions by casting out into the millennia, writing as his most distant prediction: "Intelligent beings consider the fate of the Universe." Our machines are now gods.

Six years after *Spiritual Machines*, in what Kurzweil considered an "update" of the first two books, he published the bestselling *The Singularity Is Near.* Now he adopts the concept of the Singularity and makes it his own—or, more precisely, makes it everyone's. Kurzweil is still convinced that Moore's Law will deliver us to this Promised Land, though the date has been moved out a couple decades. Now, instead of giving human beings the choice of either begging for a chance to join the great global intelligence or end up as protected zoo animals, Kurzweil is ready to let them lead the parade. At the Singularity, it won't necessarily be the computers that become conscious, but rather we humans will become the machines.

Why the shift? It may be that Kurzweil had noticed, like everyone else, that even the world's most powerful supercomputers had still not shown the slightest sign of stirring themselves to self-awareness. What he now proposed was a new definition of the Singularity, one that seemed aware of the promises (and limitations) of the Blue Brain Project and its virtual re-creation of the human brain.

Kurzweil began with a collection of premises: that the Singularity could be achieved by human beings; that because of Moore's Law it was accelerating toward us from the future; that we can understand, down to the level of neurons and electrons, how the brain functions; and that thanks to medical advances, Kurzweil's generation of baby boomers would live long enough to reach the Singularity . . . and then have a very good shot at immortality.

How would this occur? By using increasingly sophisticated tools to map the location of every neuron in the human brain and its contents and then load them into a computer where it can operate as a duplicate, virtual brain already loaded with all of our memories. This way, when the Singularity hits, we will already be aboard, Descartes's ghost in the machine, as those machines accelerate away toward their own destiny to control the universe. Thus, the race by baby boomers to stay alive until the mid-twenty-first century is not just to add a few more years to the end of a long life but a chance to become immortal, omniscient, and increasingly omnipotent.

It would be the most astonishing finish imagineable to mankind's million-year relationship with its own memories: *to become memory itself.* To reverse the equation—from our identities being defined by memory to memory (in some anonymous computer) being our identity. More than ever before, memory would truly be the guardian of all things.

BITS OF MY LIFE

There are already individuals racing to embrace this tantalizing vision. No one more than yet another computer genius, Gordon Bell, the man whose design for the Digital Equipment VAX minicomputer served as the model for the architecture of the Intel 4004, the first microprocessor.

Beginning in 1999, not long after he became a fellow at Microsoft and reaching the usual retirement age, Bell reinvented himself and embarked on a celebrated project—MylifeBits—to use the latest miniaturized, wearable digital camera, audio recorder, computer, and communications technology to document every bit of his life as he lived it. Being Gordon Bell, he also wrote the software to tie together all of these memories and archive them—a task that soon grew to a thousand photographs, several videos, hours of audio, scores of e-mails and recorded phone calls . . . per *day.* And even as he strained the capacity of his office computer's disk drive, Bell also embarked on the task of capturing and recording every surviving record of his past—from school report cards from his childhood in Kirksville, Missouri, to his founding (with wife, Gwen) of the Computer History Museum in 1979 to his receipt of the National Medal of Technology in 1991.

As Bell has said, "That is a shitload of stuff."[19]

Gordon Bell is now the most documented human being in history. As Bell claims in his book, *Total Recall,* the memory wake he will leave behind is the largest ever.[20] And yet to see Bell in person—he is a septuagenarian now—is to realize that he looks no more overburdened by his electronic appliances than the other Silicon Valley men and women sitting around him in the Stanford University coffee shop.

In a world where nearly 700 million people constantly record the fine details of their lives on Facebook, it is easy to see Gordon Bell as a pioneer of a new way of living, in which all of one's experiences live forever as a swath of artificial memory. It has been claimed that he is a kind of Kurzweil scout, cutting the path for the rest of us to the Singularity.

The self as memory, and soon, memory as self . . . it is a perfect ending to our story of human memory.

Maybe too perfect.

It may just be that Gordon Bell's destiny is not to live out Ray Kurzweil's vision, but his own. As alluring as Kurzweil's vision can be—and he has millions of ardent believers and even founded a Singularity University on the site of the old Moffett Naval Air Station in Mountain View, California (almost next door to Bell's Computer Museum)—it has not escaped considerable criticism as being as much wishful thinking as technological imperative. It may just be that Kurzweil is our Gilgamesh—a proud, accomplished man who dreams of immortality because (like billions before him) he doesn't think he deserves to die.

As for the immortality offered by Kurzweil and his notion of a Singularity, you can be sure that, should it suddenly seem imminent, there will be no shortage of volunteers: transhumanists, the mortally ill, and the adventurous. In light of this, it is hard not to be haunted by the heroic Seattle dentist Barney Clark, the man who volunteered in 1982 to be implanted with the first artificial (Jarvik) heart and who, after 112 days of confusion, misery, and the unrelenting clicking of the heart valves in his chest and head, pleaded to be allowed to die. What if the first person to wake up in a computer's memory begs to be *erased*?

Meanwhile, as Gordon Bell has always claimed for MylifeBits and his relationship to memory, it is not a revolution but an evolution; not an earthshaking transformation in the relationship between mankind and its machines but the more personal one between an individual human being and his or her memories, of the gift (or curse) of never forgetting *anything*.

As Bell has said of the experience, "It gives you kind of a feeling of cleanliness. I can offload my memory. I feel much freer about remembering something now. I've got this machine, this slave, that does it."[21]

Bell's small, tempered vision of a remembered life may not be as sweeping and apocalyptic as Kurzweil's Singularity, but it doesn't come without its own problems. The least of these is that, unlike Bell himself, most of us live pretty uneventful lives—and so terabytes upon terabytes of our life memories would probably be the worst imaginable "slides from our summer vacation" hell for others. Will even our descendants want to sift through all of this detritus of a boring life; will there be search engines to dig up the few nuggets of good stuff?

Bell's vision is also based on the assumption that we *want* to remember everything. But many people only live happy and fulfilling lives because they have managed to *forget* certain events in their past. Even a search to find and erase those memories on one's own MylifeBits might be devastatingly traumatic.

Meanwhile, as Gordon Bell knows as well as anyone, there is a certain phenomenon in the computer industry called *the legacy problem*. It is that as the years pass, computer lines tend to become less innovative because they have to pull along the burden of old programs and their loyal customers. That's precisely what happened to IBM with its 360/370 mainframe computers—and why Bell's own VAX minicomputer was so successful. Do we really want to drag along behind us *all* of the chains of the past like Jacob Marley? Or will the "clean" feeling of offloading that past into a computer be enough?

Still, there is something in Bell's vision, a thread that reaches back through the history of mankind to a dream even older than that of immortality. It is for one's brief time on this Earth to have meaning, for it to echo down through history if only as the faintest memory. It is the oldest human voice on earth whispering, *Don't forget me.*

MEMORY LOSS

The Rosicrucian Egyptian Museum, constructed to resemble the ancient Temple of Amon at Karnak, has stood in San Jose, California, for seventy-five years. It is a mile from the warehouse where Rey Johnson and his team built the first disk drive, three miles in different directions from where Al Shugart led the creation of the minidisk drive and Steve Jobs and Apple prototyped the iPod. Another couple miles and you reach the laboratory (now a retail store) where Bob Noyce and the Traitorous Eight invented the integrated circuit. Head from there a couple miles toward San Francisco Bay and you reach the site of Fairchild, where Gordon Moore devised his Law, and a quick hop across the freeway from there and you arrive at Ray Kurzweil's Singularity University. Like memory itself, most of these places and events are long gone, yet in remembering them, they are still in the present.

The museum is run by the Ancient and Mystical Order Rosae Crucis—the Rose and the Cross—Rosicrucians. The AMORC, though founded in 1915, claims roots dating back to ancient Egypt. Claiming among its past members Francis Bacon, René Descartes, and other figures who have appeared in this story of memory, the Rosicrucians are yet another surviving branch of the occult/secret-knowledge/syncretic belief system that we have seen wax and wane over the millennia.

These days, other than making for some entertaining conspiracy theories, the Rosicrucians, like other mystical groups, have retreated to tiny, self-nurturing communities waiting for the world to turn once again. They are a reminder that to be remembered is to endure.

In one of the exhibit cases at the museum, amidst the mummies, canopic jars, and exquisite lapis lazuli jewelry, is a small coffin bearing the large-eyed, high-checkboned face of a young girl. This coffin is estimated to be almost 2,600 years old—and from its hieroglyphics can be deciphered a name: Ta'awa.

Other than that she came from a wealthy family, her name—and, in fact, just her nickname—is all that we know, and will likely ever know, about Ta'awa. But that is enough. She is still remembered after 2,500 years, in a world she could not have imagined. And her name has survived because it was written on a board inside a buried coffin that, though slightly browned, probably looks as fresh as the day it was made. So, too, do the names on the papyrus fragments in the display cases nearby.

So Ta'awa, despite her brief life, has found her own form of immortality. So has King Gilgamesh, who lives on twelve clay tablets, baked to stone, that rest in the British Museum. Fifty miles away from there in Oxford, in the dark old Duke Humfrey's Room of the Bodleian Library, the great bestiary, *Bodley 764,* slowly breathes its way through the centuries, its vellum pages still supple, the colors and gilding of its paintings bright and new.

Israeli-born Arik Paran, who lives in nearby Sunnyvale, knows the Egyptian Museum well. When his three boys were young, touring the Rosicrucian Museum was always an annual elementary-school field trip. And like every other visitor, even though he had seen the comparable antiquities of his own country, Paran was astounded by the sheer age of the artifacts on view.

These days he appreciates their durability more than ever. An engineer by training, a few years ago Paran caught the entrepreneurial bug, quit his job, and founded his own company in San Francisco, Digital Pickle, dedicated to the restoration of old audio and video recordings, as well as computer files; converting obsolete memory media to state-of-the-art new forms. It was a nice little business: Private and corporate customers would bring in old videocassettes, floppy disks, 8-mm films, professional videotapes, microcassettes, and all sorts of other once-popular memory storage media. Paran had the equipment in house to capture stored data, "sweeten" it by adjusting the contrast, heightening the color, pulling voices up through the tape hiss, merging scores of small files on multiple floppies . . . and then downloading the results onto single DVDs.

It didn't take long for Paran to realize just how *poor* the quality of these recordings was. Videocassettes created and sold just a decade ago had already begun to bleach out; voices and songs on cassettes had begun to fade away, and the Mylar plastic in the floppy disks had begun to crack or drop bits of memory. But most distressing was working with the big one-inch and two-inch videotape, the kind used in studios to record important, professional-quality videos. Sometimes Paran or one of his staff would thread the tape into its player and see only static on the screen. They would jump to shut down the player, but inevitably some of the images would be lost forever. And opening the player, Arik would check the read-write head to find that the old magnetic surface layer on the tape had literally peeled off like old paint attacked by a scraper. Sometimes the rest of the surface could be affixed to the tape, but just as often the tape couldn't be salvaged. He tried not to think about what had been lost.

LOST SOULS

For two thousand years, after parchment and/or rag paper became the artificial memory of choice for China, the Middle East, Europe, and then the rest of the world, scarcity, not preservation, has been the biggest challenge. Once a book was written or printed and put on a library shelf, its life expectancy was measured in centuries.

But all of that ended around the time of Herman Hollerith and his punched cards. Paper tape, made fragile lace with thousands of punched holes, wasn't designed to last more than a few weeks—just long enough to transfer its data. Early celluloid film stock, as we all know, was terribly volatile, even explosive. An estimated half of all silent films are lost forever, which is why the discovery in South America of missing reels of *Metropolis* or rumors of the rediscovery of *London After Midnight* reverberate around the world. The few pioneering Edison films survive mostly because the company made test copies of the frames on paper.

Meanwhile, even early films from the sound era have begun to fade, especially those that used experimental color techniques. The most popular solution—that of creating a "master" that can be carefully preserved—merely shifts the problem to that of the danger of having only one of a kind—as the infamous 1967 MGM movie vault fire, which led to the loss of hundreds of films—shows.

Printing, traditionally the most reliably durable form of mass artificial memory, went through its own transformation, too. With the number of avid readers growing into the millions, books, like newspapers before them, shifted to a lower-cost production medium: wood pulp paper. Cheap and abundant, pulp paper fueled the middlebrow home library boom of the first half of the twentieth century and the paperback boom of the second half. But pulp paper is highly vulnerable to heat and light, as any owner of a brown-papered, crumbling old paperback or newspaper knows.

Magnetic memory, when it appeared, was hailed not just for its breakthrough in capacity, but also its durability. After all, audio and video tape was made from the newest space-age material, which seemed infinitely tougher than anything that had come before. As for hard-disk drives, they were milled out of solid metal and coated with rust—what could be more elemental? Well, actually, silicon semiconductors.

These memory devices were thought to be all but immortal, the first truly worthy replacement for the book. A few decades later, we know better. What is easily erased can usually be easily erased forever. Read-write heads skimming at breakneck speed over the surface of a disk can lose direction and auger in, gouging the oxide like a plow. Accidentally (or purposefully) create a powerful magnetic field and all of those little molecules will align in a different way and forget everything they knew. Silicon chips are tougher, but the lead frames in which they are implanted aren't. And leave a motherboard stuffed with memory chips running for too long with insufficient cooling and those chips will burn themselves out one by one. Did I mention electrical surges?

And then, of course, there is the potential for catastrophic loss. What if the sun gets testy and the Earth undergoes a solar storm like it did in 1859—what is called the Carrington Event? In that largely pre-electric world, there were enough charged particles emitted by the sun to cover the planet with an aurora borealis bright enough to read by and to cause telegraph equipment to burst into flames.[22]

If the Earth was to be bombarded now in the same way, it could erase much of the memory that now exists in magnetic storage—not to mention batteries and the power grid, and the electric motors and integrated circuits that power them. So, even if your disk survives, it may take you a long time to find something with which to access it. And just imagine—post-Singularity—if you were to be living inside one of those computers. . . .

Meanwhile, as all of your magnetic memory erases itself under this onslaught, the world's books, quietly resting on shelves in private dens and public libraries (at least those that survived de-accessioning and pulping to make room for computers), will be undisturbed other than by the flickering lights and the angry shouts of hobos doing online gambling.

Of course, there are always lasers. CDs and DVDs were initially promoted as lasting almost forever. All of us know by experience that isn't true; in fact, a scratch in the wrong place on a CD is often more catastrophic than one on the LP record it was designed to replace. Still, the CD and DVD are more resistant to magnets because they operate from a spiral of up to a couple billion pits just beneath their polished surface. A low-power laser reads those pits and converts them to sound or video. But heat, light, and various forms of radiation still affect these various versions of the optical disk, and while manufacturers claim that these disks—read-only and read-write, low-density or Blu-ray—should last as long as fifty years, more objective testers put their life expectancy at half that . . . and some CD-Rs have degraded after less than two years.

Asks Christopher Mims of *Technology Review*:

> *It's tempting to believe that we live in a special time—this is the root of all apocalyptic thinking—but it's hard to compare even today's*

> *menaces to the rise of the Third Reich, the fall of the Roman Empire or the Black Death. At least not yet.*
>
> *But supposing something were to happen, as it does every day in parts of war-torn sub-Saharan Africa—some cascade of environmental and political disasters leading to armed conflict or resource starvation. What happens when all those data centers, housing all that knowledge we digitized without a second thought, go dark?*[23]

If the story of memory teaches us anything, it is that if you wait long enough, the worst *will* happen. Those worst-case scenarios for humanity Mims presents have only happened in just the last two thousand years of mankind's 200,000-year history. And only a fool believes they won't happen again; only an idiot doesn't prepare for their arrival.

It is often said that civilization depends upon each generation assuming its responsibilities and then passing them on to the next. Whether we realize it or not, memory, at least in the digital age, appears to require the same commitment. We are like the Romans, enjoying the new communications revolution wrought by papyrus—but also recognizing that we must constantly copy and update our fragile scrolls or they will be lost forever. Already, some of the early months of the World Wide Web are lost forever because no one made screen grabs and copies.

As exciting or terrifying as the idea of memory implants, life recording, and the Singularity may be, none of them will ever take place if this fourth scenario—*forgetting*—arrives first. And if it does, we may get a closer look at the eighth century than we'd like.

We may never want the kind of immortality that requires becoming one with our computers. But for the first time in history, we have the chance to have the memories of all of our lives live on indefinitely after us, to leave wakes in time as great as those once made only by kings. But it will only happen if we don't forget to remember, to protect the record of our time in this world, and most of all, to find new, more enduring ways to preserve our memories.

Memory is the guardian of all things. But in the end, we are the guardians of memory.

Acknowledgments

This book has been germinating in my own memory for many years. In fact, looking back, I think that I first became interested in the concept of human memory while I was still in elementary school, sick at home with the flu and watching memory prodigies showing off their skills on *Jeopardy!* and the *GE College Bowl.* That was almost a half-century ago now. And I can remember taking a computer programming course with my father (whose story begins this book) a few years later at NASA's Ames Research Center and spending hours typing and collating punch cards—and hearing the instructor describe the new "drum" memory that would soon make all of this drudgery obsolete.

But perhaps my first real appreciation of the endless power of human ingenuity to create new means of storing memories came just a few years later, as a high-school senior interning in the Ames Research Center, when for the first time, I downloaded the results of my work onto a magnetic card on my Hewlett-Packard 9800 desktop calculator. That may have been the moment when I first start thinking seriously about memory. And that interest was driven home a few years later when I first watched James Burke's BBC miniseries *Connections* and its legendary opening about what would happen if the modern world were to suddenly stop and we had nothing left but our memories to lead us forward.

Life has an odd way of circling back on itself. Within the next two decades I would find myself as a corporate publicist at Hewlett-Packard helping to introduce the second generation of programmable handheld calculators, then as a technology reporter covering the rise of floppy and Winchester disk drives, and even filming public television shows with the extraordinary Mr. Burke. Through it all, I never ceased to be in awe of the men and women who made the memory revolution. They have never received the credit they deserve for keeping the electronics age alive over the last seventy years. Their work is the true miracle of our technology era, and though their names are too many to fit into this section, I want to once again acknowledge their historic achievement.

Among the true legends of the memory revolution, I want to particularly thank two gentlemen who through the years were always there to generously answer my questions, correct my errors, and at the beginning, work patiently with a cub reporter: Dr. Gordon Moore and the late Al Shugart Jr. Gordon's reputation (and his law) are so synonymous with the electronics age that I assume his name will never be forgotten. I'm not so sure that will be true for Al Shugart—which is a pity, because not only was he one of the great inventors of the twentieth century, but no one ever seemed to have so much fun in tech as he did. We in Silicon Valley could use a lot more of Al's spirit and sense of play.

The actual idea for this book came, ironically, over a couple lunches and dinners, marinated with too many bottles of wine, with James Burke, the man who brought the playfulness and excitement of science history to many of my generation—and who proved to be exactly the wonderful dinner companion we all imagined him to be. As I wrote this book, I often thought of James, enjoying retirement in France, and tried to live up to the high standards of his storytelling.

I want to acknowledge the influence of others as well. Steve Forbes, Tim Forbes, and Rich Karlgaard, in letting me run *Forbes ASAP* with complete editorial independence, enabled me to begin chasing down the back alleys of tech history to find this story. Before that, Jim Mitchell, my editor at the *San Jose Mercury-News,* with whom I famously battled, trusted my news judgment enough to let me write the very first newspaper stories about the memory industry. Gordon Bell, my old neighbor, whom I've watched through several incarnations, both inspired me with his willingness to put his own life on the line for his vision, and gave me the ending to this book. And Ray Kurzweill, who often wrote for me at *Forbes,* and whose Singularity Institute is just down the road from where I'm writing this, provoked my thinking about the future convergence of human and artificial memory more than anyone else.

For the Bestiary section of this book, I want to thank the past and present librarians of the Bodleian Library at Oxford, Reg Carr and Sarah Thomas, the director of special collections Richard Ovendon, and Martin Kauffmann, curator of medieval manuscripts. I am truly appreciative of their willingness to take the time to talk with me, dream with me, and most of all, introduce me to the unforgettable *Bodley 764.*

For the final Egyptian section of the book, I want to acknowledge the assistance of guide Terrance Gamble, who took me through the Rosicrucian Museum to find just the right historic object to describe, and story to tell.

As always, through each of my score of books to date, my wife, Carol, has been the rock on which my odd career rests. Once again, even while managing her own, now international, career as a painter, she made sure the bills were paid, the kids fed, and our family healthy. And, once again, her assistant, Leslie Johnson Lopez, was always there to help me with the details—computers, mailings, contacts, etc.—that go with a writer's life. My old TV production partner, Robert Grove, helped flesh out sections in the earliest outline of this book.

In the end, the only reason the book you hold in your hand even exists is because of two people. One of the lessons of Silicon Valley is to always work with people more talented than you. And a dozen years ago I found that person in Jim Levine. In the world of business books, Jim is already known as a superstar literary agent, but what is rarely acknowledged is Jim's unparalleled ability to find the real story in a book idea. Consider that, in the last five years, I've come to Jim with unlikely notions about a radically new model of business organization, a biography of two dead business titans, and a book about the history of memory, and Jim not only found the heart of each of them, but found major publishers for them as well.

That said, I know Jim will also agree with me that it takes a very special editor to see the potential of a book with as unlikely a subject as this one. But Phil Revzin, and later Nichole Argyres, did just that, to my eternal gratitude. And, I'm afraid to say, Nichole and her assistant, Laura

Chasen, had to show more patience with me—as Silicon Valley swept me up again in its latest boom—than any of my editors before them. For that I hope they accept my apologies.

To all those individuals I just mentioned, and to everyone else who helped in big ways and small to make this book possible, my thanks. You will always be in my memory of *The Guardian of All Things*.

Notes

1. Finding a Voice

1. Robin McKie, "How a Hobbit Is Rewriting the History of the Human Race," *The (Guardian) Observer*, February 21, 2010.
2. http://sjohn30.tripod.com/id1.html.
3. http://anthropology.net/2007/10/18/neandertals-have-the-same-mutations-in-foxp2-the-language-gene-as-modern-humans/.
4. www.andreasbick.de/en/writings/sound_reading/?article=111.
5. "Toba Catastrophe Theory," *ScienceDaily.com*, www.sciencedaily.com/articles/t/toba_catastrophe_theory.htm.
6. J. David Sweatt, "The Neuronal MAP Kinase Cascade: A Biochemical Signal Integration System Subserving Synaptic Plasticity and Memory," *Journal of Neurochemistry* 76 (2001): 1–10.
7. www.enotes.com/topic/Baddeley<#213>s_model_of_working_memory.
8. www.audiblox.com/human_memory.htm.
9. www.longtermpotentiation.com. Use the March 1, 2011, entry for an excellent explanation.
10. Endel Tulving, "What Is Episodic Memory?" *Current Directions*, 1993, available online at http://alicekim.ca/ET93.pdf.
11. www.activemind.com/Mysterious/Topics/SETI/drake_equation.html.

2. The Cave of Illumination

1. http://communications.uvic.ca/releases/tip.php?date=22022010.
2. http://findarticles.com/p/articles/mi_hb3284/is_295_77/ai_n28995839/.
3. www.omniglot.com/writing/vinca.htm.
4. www.ancientscripts.com/indus.html.
5. www.feelnubia.com/index.php/culture/tongues/133-the-ancient-nsibidi-writing-system.html.
6. http://incas.homestead.com/quipu/caral_oldest_quipu.html.

7. "About Sequoyah," http://www.sequoyahmuseum.org/index.cfm/m/1/fuseAction/contentpage.main/detailID/29.

3. Clay, Reeds, and Skin

1. Martin Litchfield West, *The East Face of Helicon: West Asiatic Elements in Greek Poetry and Myth* (Oxford, UK: Oxford University Press, 1997), 334–402.
2. www.ancienttexts.org/library/mesopotamian/gilgamesh/.
3. www.buzzle.com/articles/history-of-egyptian-hieroglyphics.html.
4. Rosicrucian Museum, San Jose, California.
5. www.buzzle.com/articles/history-of-egyptian-hieroglyphics.html.
6. http://legacy.earlham.edu/~seidti/iam/papyrus.html.
7. Robin McKie, "Arab Scholar 'Cracked Rosetta Code' 800 Years Before the West," *The (Guardian) Observer,* October 3, 2004.
8. www.rosettastonelanguagekey.com/html/an_ancient_mystery/egyptian_hieroglyphs.htm.
9. www.crystalinks.com/libraryofalexandria.html.
10. Ibid.
11. www.alpharubicon.com/primitive/tanningdragoona.htm.
12. http://en.wikipedia.org/wiki/Parchment.
13. http://elab.eserver.org/hfl0243.html.
14. http://cunnan.sca.org.au/wiki/Codex.

4. The Bloody Statue

1. Dr. Charles Fernybough, "Moonwalking with Simonides" review, *Psychology Today* blog, April 23, 2011, www.psychologytoday.com/blog/the-child-in-time/201104/moonwalking-simonides.
2. www.iep.utm.edu/cicero/.
3. http://public.wsu.edu/~dee/REN/HUMANISM.HTM.
4. Cicero, *De Oratore* II (lxxxvi), 351–54.
5. *Rhetorica ad Herennium,* Book III, Chapter 22.
6. Ibid., Book III, Chapter 19.
7. Joshua Foer, *Moonwalking with Einstein* (New York: Penguin Press, 2011), 248.
8. www.iep.utm.edu/cicero/.
9. www.roman-empire.net/articles/article-003.html.
10. http://en.wikipedia.org/wiki/Michael_Psellos.
11. Richard Erdoes, *1000 AD* (Berkeley, CA: Seastone, 1998), 60–61.
12. http://everything2.com/title/Historical+Evidence+Regarding+the+Libraries+of+Muslim+Spain.
13. Erdoes, *1000 AD*, 60–61.
14. Karl Christ, *Handbook of Medieval Library History* (New York: Scarecrow, 1984), 14–15.
15. "Celebrities in the History of Printing," www.chinaculture.org/library/2008-02/06/content_46431.htm.

5. Long-Leggedy Beasties

1. Charles Homer Haskins, *The Renaissance of the Twelfth Century* (Cambridge, MA: Harvard University Press, 1927), viii.
2. Ibid., 6–8.

3. Lawrence M. Principe, *The Scientific Revolution* (Oxford, UK: Oxford University Press 2011), 7.
4. www.cosmopolis.com/villa/liberal-arts.html.
5. www.chinaculture.org/gb/en_aboutchina/2003-09/24/content_26624.htm.
6. T. H. White, *The Book of Beasts* (Mineola, NY: Dover, 2010).
7. George McCauley Trevelyan, *A Shortened History of England* (New York: Penguin, 1988), 69.
8. White, *The Book of Beasts*, 240.
9. Ibid., 241.
10. Ibid., 244.
11. Ibid., 5.
12. Author's conversation with Martin Kaufmann, November 2008.

6. Theaters of Memory

1. Michael Lewis, "The Roasting of Giordano Bruno," *Slate*, February 28, 2000, www.slate.com/articles/news_and-politics/i-see-france/2000/02/the-roasting-of-giordano-bruno.html.
2. http://bigbendnow.com/2011/02/giordano-bruno-martyr-or-fool/.
3. http://romaexperience.com/romediary/giordano-bruno/.
4. Frances Yates, *The Art of Memory* (Chicago: University of Chicago Press, 1966), 12.
5. http://galileo.rice.edu/chr/bruno.html.
6. Yates, *The Art of Memory*, 41.
7. Ibid., 130–31.
8. Ibid., 131–32.
9. This explanation taken from the Church of St. James the Great's booklet on its murals.
10. William Boulting, *Giordano Bruno: His Life, Thought, and Martyrdom* (New York: E. P. Dutton 1916), 58.
11. *Select Charters and Other Illustrations of English Constitutional History*, ed. W. Stubbs, 2 vols., 9th edition (London: Clarendon Press, 1913), 176.
12. M. T. Clanchy, *From Memory to Written Record, England 1066–1307,* 2nd edition (Oxford, UK: Blackwell, 1993), 115.
13. Ibid.
14. www.greatsite.com/timeline-english-bible-history/william-tyndale.html.
15. www.tititudorancea.com/z/encyclopedia_35892.htm.

7. Patterns in the Carpet

1. Alfred Chapuis and Edouard Gélis, *Le Monde des automates,* volume 2 (Paris: Neuchatel 1927), 149–51; Alfred Chapuis and Edmond Droz, *Automata* (Paris: Neuchatel, 1958), 233–34; Jessica Riskin, "The Defecating Duck, or, the Ambiguous Origins of Artificial Life," *Critical Inquiry* (September 2003).
2. www.antikythera-mechanism.gr/.
3. "Seventh Olympic Ode" by Pindar, www.jstor.org/pss/4430612.
4. http://library.thinkquest.org/C006011/english/sites/heron_bio.php3?v=2.
5. Joseph Needham, *Science and Civilization in China: Volume 2. England* (Cambridge, UK, Cambridge University Press, 1986), 53.
6. Ibid.
7. Ibid., 54.

8. www.classicallibrary.org/descartes/meditations/.
9. Riskin, "The Defecating Duck."
10. Ibid.
11. www.newadvent.org/cathen/10325a.htm.
12. http://research.miralab.unige.ch/automata/eightennth/vaucanson_uk.htm.
13. Book review, "Living Dolls: A Magical History of the Quest for Mechanical Life," by Gaby Wood, *The Guardian*, February 16, 2002.
14. Ibid.
15. Ibid. For a good history of automatons, including the intriguing fate of the Digesting Duck (which may still exist), see *Living Dolls: A Magical History of the Quest for Mechanical Life* by Gaby Wood (Faber & Faber, 2003).

8. Tick, Talk

1. http://wiki.whitneygen.org/wrg/index.php/Archive:Domesday_Book.
2. Adam Goodheart, "The Census of Doom," *New York Times*, April 1, 2011, online edition, http://opinionator.blogs.nytimes.com/2011/04/01/the-census-of-doom/.
3. Ibid.
4. Ibid.
5. Ibid.
6. Ibid.
7. William Aul, "Herman Hollerith: Data Processing Pioneer," *Think* (IBM employee magazine), November 1972, 22–24. Available at www-03.ibm.com/ibm/history/exhibits/builders/builders_hollerith.html
8. Ibid.
9. Ibid.
10. Michael S. Malone, *The Big Score* (New York: Doubleday, 1985), 14.
11. Nikola Tesla, "Thomas Edison," *New York Times*, October 19, 1931.
12. "George Eastman," www.nndb.com/people/980/000086722/.
13. Quoted in Patrick Robertson's *Film Facts* (New York: Billboard Books, 2001), 5.
14. www.acmi.net.au/AIC/MAGIC_MACHINES_4.html.
15. Some material in this section is from www.ce.org/Events/Awards/468.htm.
16. http://cs-exhibitions.uni-klu.ac.at/index.php?id=220.
17. http://hyperphysics.phy-astr.gsu.edu/hbase/audio/bias.html.
18. www.videointerchange.com/wire_recorder1.htm.
19. www.computerhistory.org/events/.
20. Malone, *The Big Score,* 67.

9. Diamonds and Rust

1. www.cci-compeng.com/Unit_5_PC_Architecture/5202_Who_Invented.htm.
2. Lawrence M. Fisher, "Reynold Johnson, 92, Pioneer in Computer Hard Disk Drives," *New York Times,* September 18, 1998.
3. Dinner talk by Rey Johnson at the DataStorage '89 Conference, September 19, 1989. Available at www.magneticdiskheritagecenter.org/100th/reyjohnson.htm.
4. George Rostky, "Disk Drives Take an Eventful Spin," *EE Times,* July 13, 1998.
5. www-03.ibm.com/ibm/history/exhibits/storage/storage_3340.html.
6. Ibid.
7. Malone, *The Big Score,* 279.
8. Michael S. Malone, *Betting It All* (Wiley, 2002), 129–30.

9. Ibid., 129.
10. Ibid., 131.
11. Ibid.
12. Rostky, "Disk Drives Take an Eventful Turn."
13. Malone, *Betting It All*, 133.
14. Ibid., 136.
15. Ibid.
16. Ibid.
17. www.seattlepi.com/default/article/Apple-s-new-iPod-player-puts-1-000-songs-in-your-1070406.php.
18. www.cedmagic.com/history/transistor-1947.html.
19. Michael S. Malone, *The Microprocessor: A Biography* (New York: Springer, 1995), 33.
20. Ibid., 53.
21. www.computerhistory.org/semiconductor/timeline/1958-Miniaturized.html.
22. Malone, *The Microprocessor,* 54–55.
23. Ibid., 56–58.
24. http://arstechnica.com/hardware/news/2008/09/moore.ars.
25. Michael S. Malone, "The Future Still Lives," *Forbes*, May 11, 2011.
26. Malone, *The Microprocessor*, 3–15.
27. http://en.wikipedia.org/wiki/Programmable_read-only_memory.
28. http://searchcio-midmarket.techtarget.com/definition/EEPROM.
29. http://electronics.howstuffworks.com/flash-memory.htm.

10. The Persistence of Memory

1. www.wisegeek.com/how-big-is-the-internet.htm.
2. Eliphas Lévi, *Transcendental Magic: Its Doctrine and Ritual* (The Occult Publishing House, 1860, 1913), 108.
3. http://psychology.about.com/od/biopsychology/ss/brainstructure_2.htm.
4. www.musanim.com/miller1956/.
5. Henry Markram, from 2009 TED Conference, Oxford University, http://www.youtube.com/watch?v=LS3wMC2BpxU.
6. Jonathon Fildes, "Artificial Brain '10 Years Away,'" BBC News online, July 22, 2009, accessed February 21, 2012, news.bbc.co.uk/2/hi/8164060.stm.
7. H. G. Wells, *World Brain* (New York: Doubleday), xvi; George B. Dyson, *Among the Machines* (New York: Basic Books, 1997).
8. Ibid., 87.
9. Garet Garrett, *Ouroboros* (New York: E. P. Dutton, 1926), 19.
10. Dyson, *Darwin Among the Machines*, 227.
11. Ibid., 1–2.
12. Michael Chorost, *World Wide Mind* (New York: Free Press, 2011), 9–10.
13. Dr. David Whitehouse, "Computer Uses Cat's Brain to See," BBC News Online, October 8, 1999, accessed February 21, 2012, news.bbc.co.uk./2/hi/sci/tech/468857.stm.
14. Ian Sample, "The Brain Scan That Can Read People's Intentions," *The Guardian,* February 9, 2007.
15. www.singularity-universe.com/technologicalsingularity. Primary source for Irving Good quote: www.acceleratingfuture.com/pages/ultraintelligentmachine.html.
16. www-rohan.sdsu.edu/faculty/vinge/misc/singularity.html.

17. http://reason.com/archives/2007/05/04/superhuman-imagination.
18. Ibid.
19. Clive Thompson, "A Head for Detail," *Wired*, November 1, 2006.
20. Gordon Bell and Jim Gemmall, *Total Recall: How the E-Memory Revolution Will Change Everything* (New York: E. P. Dutton, 2009).
21. Thompson, "A Head for Detail."
22. http://news.nationalgeographic.com/news/2011/03/110302-solar-flares-sun-storms-earth-danger-carrington-event-science/.
23. www.technologyreview.com/blog/mimssbits/26889/?p1=blogs.

Index